# THE BODY

*The Body* is a multidisciplinary collection of essays from a number of world-renowned experts on the topic of the body arising from the 14th Annual Darwin College Lecture Series. It deals with our understanding of the modern body from a number of different standpoints. The opening chapters treat developmental biology of the body, the Human Genome Project, and *in vitro* fertilisation and the possibility of human clones. Other chapters explore the subject of bodies and the criminal mind, dead bodies and human rights, science and the politics of incarnation, bodies at the boundaries of pornography and art, and the body in archaeology.

THE DARWIN COLLEGE LECTURES

# THE BODY

Edited by *Sean T. Sweeney* and *Ian Hodder*

CAMBRIDGE
UNIVERSITY PRESS

PUBLISHED BY THE PRESS SYNDICATE OF THE UNIVERSITY OF CAMBRIDGE
The Pitt Building, Trumpington Street, Cambridge, United Kingdom

CAMBRIDGE UNIVERSITY PRESS
The Edinburgh Building, Cambridge CB2 2RU, UK
40 West 20th Street, New York, NY 10011–4211, USA
477 Williamstown Road, Port Melbourne VIC 3207, Australia
Ruiz de Alarcón 13, 28014 Madrid, Spain
Dock House, The Waterfront, Cape Town 8001, South Africa

http://www.cambridge.org

First published 2002

Printed in the United Kingdom at the University Press, Cambridge

Typeset in 10/14 IronSB QuarkXpress [wv]

A catalogue record for this book is available from the British Library

Library of Congress Cataloguing in Publication data
The body / edited by Sean T. Sweeney and Ian Hodder.
    p. cm. – (The Darwin College lectures)
  Includes bibliographical references and index.
  ISBN 0 521 78292 9
    1. Body, Human – Social aspects. 2. Human physiology. 3. Human genome.
4. Human rights. I. Sweeney, Sean T., 1966–   II. Hodder, Ian. III. Series.

HM636.B58 2002
306.4 – dc21   2001052877

ISBN 0 521 78292 9 hardback

# Contents

The plate section is between pp. 26 and 27.

# Introduction

SEAN T. SWEENEY AND IAN HODDER

## Why the body is of topical interest

Many attempts have been made in recent years to make sense of the increased awareness of the body that we see around us, both in society and popular culture and in scholarly disciplines. The following summary of these attempts at interpretation of the rise of interest in the body responds in particular to the writings of Bryan Turner, Simon Williams and Gillian Bendelow, and Anthony Giddens.

Many of our personal concerns about our bodies relate in some way to consumerism. Turner argues that the current fascination with the body is related to a shift from 'the labouring body' to 'the desiring body'. Taking a long-term view, feudal and industrial bodies were tied to property, ownership and control. In the post-industrial age the body has become separated from the economic and political structure of society. Contemporary consumerism sees the body as the locale for pleasure, desire, playfulness. Unemployment, early retirement and the rise of leisure and service industries all conspire to focus on the body as consumer product to be manipulated and packaged. This is the body as sign. The body is pampered, groomed, massaged, exercised, tanned, pierced, tattooed. In contemporary consumer society, diet is a central part of looking good. The emphasis on achieving personal happiness through having a beautiful body has led to slimming, workouts – and it has also led to a rise in anorexia and bulimia (for more individual and psychological reasons for these, see Canter, Chapter 4). To be beautiful and happy you also have to look youthful – hence facial surgery, implants, hair dying, and cellulite removal.

Consumerism, but also the wide availability of contraception and divorce, have transformed interpersonal relations. Giddens has claimed that in

Western capitalist societies we are seeing a transformation of sexual relationships. Simple, standardised and oppositional codes are being transformed, especially by gay movements and by women, into a 'plastic' sexuality – one that can be modelled to personality and lifestyle. These new relationships may be more equal, and more contingent (rather than being 'forever'). All this has led to a renewed interest in embodiment in the sense of bodily and sexual identity. The new forms of relationships have their own tensions and insecurities that often lead to therapies that again look inward at mind and body. As we 'get in touch with' ourselves, so the issue of our bodies comes to the fore. Contemporary bodies are increasingly tied to self-understanding, intimacy, self-realisation and personal fulfilment. There is thus an inward-looking focus on the body, and a wide range of mystical, New Age, alternative medicine, etc. perspectives on the body have emerged. This is partly in response to the intrusion of consumerism and the state into our lives – we try to control at least our own bodies, even if we cannot control anything else.

The Women's Movements and various forms of feminism have turned attention to the body as part of a wider critique and overturning of male control and objectification. Much feminist work deals with sexual 'difference' and with the ways in which women's bodies are and have been manipulated, dominated and oppressed. Sex, gender and subjectivity are seen as constructed within systems of power and knowledge. The historical 'making' of sexed bodies has been charted from the Classical Greeks to Sigmund Freud. One masculinist response has been to re-assert the male body. Masculinity has recently been re-theorised to focus on the differences between dominant and alternative masculinities, on the dependence between homophobia and heterosexual masculinities, and on heterosexual masculinity as a hegemonic discourse.

Another important influence on our understanding of the body in the twenty-first century has been photography. The role of photography of the body in mass killing and war is profound. Several of the chapters in this volume refer directly or indirectly to images of the First World War trenches, or of the burning body of the little Vietnamese girl running down the road, of the bodies of Bosnian and Kosovar men and women in mass graves, and above all of piles of naked corpses in the Nazi concentration camps. These images of naked power crushing naked frailty have dominated the twentieth and twenty-first centuries. They are an essential component of our repulsion at our history and of our preoccupation with bodies. In these images the body becomes a symbol of freedom, individuality and meaning in opposition to the state (see Canter, Chapter 4).

An intellectual effect of the holocaust was a fear of racism and biological reductionism – any approach that assumed that biological characteristics could be linked to qualities of intelligence, social abilities and meaning. The measuring of skulls and bodies to produce racial 'types', a practice common in the late nineteenth and earlier twentieth centuries, was decried as reductionist and dangerous. The notion that biology determined behaviour was eschewed.

Molecular biological research has opened an interest in the biological within us. Without returning to the excesses of biological determinism and reductionism (though some have tried, in, for example the search for genes determining sexual orientation, see Latour, Chapter 7), there is a concern to explore (as in twins studies) the interactions between genes and culture, between our genetic inheritance and our environment. There is a move to enhance the quality of life by understanding the genes that control hereditary diseases. Controversially, some hereditary diseases are selected against by using pre-implantation screening of embryos. This opens a debate about the nature of 'biological destiny', what qualifies as a disease, which diseases should be selected against by using such procedures, and issues of 'quality of life'.

Already in traditional photography there was increased awareness that the body could be manipulated and 'set up' to look different, better, more beautiful in different contexts. The dispersal of bodies has since become ever more prevalent with digital technologies and the internet. Not only can images of bodies now be air-brushed to remove blemishes, but they can be cut up and dispersed on the internet. Images of sliced and dissected bodies can be explored on the internet, and students can engage in virtual medicine. We can indulge in cybersex and reveal only those parts of our selves we want to reveal.

New medical technologies have also played a major role. Microsurgery now makes trans-sexualism an option – there can be an emphasis on the changing, unfixed character of bodies. IVF (in vitro fertilisation), heart and other organ transplants, cybernetics, new reproductive technologies, cloning – all these have raised major ethical questions about what is 'a body', who 'owns' parts of bodies, when does life begin, when does it end. The AIDs epidemic and reproductive choice arguments have concentrated debate about individual rights and liberties.

Shifting demographic patterns have also been significant. Chronic illness in an increasingly aged population has focused attention on the long-term care of the body, its ageing, health, and the boundaries with death (euthanasia).

For all these reasons, we have what Turner calls 'the rise of the somatic society': this is 'a society in which our major political and moral problems are expressed through the conduit of the body'. These are also the reasons why we felt that a book on the body was timely. Advances in our understanding of the body are being made in many different fields, stretching from the arts and humanities, through the social sciences, to the biological sciences. By bringing representatives of these different fields together, we have tried to open up debate and to search for common themes that resonate with wider public worries about the body and about the intrusion of science into the body.

### One body: history of research trying to erode the culture/biology, mind/body splits

The early Cartesian view of the body that dominated rational and Enlightenment thought was that the mind was constrained by the excretions and emotions of the body. It had to be released from such concerns. The natural bodily forces had to be tamed, labelled and disciplined, so that contemplative and higher thought could be achieved. The rational was elevated over the corporeal. Mind was thus separated from body. In Descartes' famous phrase: 'cogito ergo sum' (I think, therefore I am), biology was separated from culture.

Much social theorising about the body in the twentieth and twenty-first centuries has tried to escape this dualistic view of the body. *The Body* tries to bring the sciences and humanities together in a consideration of the body as a singular whole.

In classical sociology and anthropology we can see the gradual shift towards a non-dualistic view. Anthropologists such as Mary Douglas saw the body as an important and potent symbol, expressing divisions such as sacred/profane, purity and danger. The different ways in which the body itself is experienced in different cultures have been explored. In sociology, the classical view focused on a rational, disembodied, decision-making process bound by utility and least-cost criteria. There was a fear of biological reductionism, racism and sexism. This meant a lack of discussion of biology, bodies and embodiment in favour of distanced accounts of rational behaviour.

For Karl Marx, the biological body depended on nature, but through definite social relations and social praxis. Thus humans make themselves in an active, embodied and sensuous way. Friedrich Engels explored the way in which bodies of different classes lived to different ages and had differing health. In many disciplines, however, these Marxist ideas became subsumed within a materialism in which the body was constructed by the material forces of production. In these developments, the body became a less lived and more objective product.

Again taking a historical perspective, Max Weber saw the rise of rational management and control of the body and its emotions linked to the Protestant ethic and the rise of capitalism. According to the Protestant ethic, pleasures of the flesh were to be denied in an ascetic body engaged in relentless labour. Weber's focus on the control and surveillance of bodies in capitalism converged with Michel Foucault's work on disciplinary technologies. Foucault describes the rise in the late eighteenth and nineteenth centuries of institutions for the disciplining and surveillance of the body. He sees this in clinical investigation of the body, but also in schools, factories, military academies, in discourse on sexuality, and in prisons. Foucault's work focuses on the social construction of bodies as part of power/knowledge.

More recently there has been much new work that attempts to deal with the opposition between an organic 'pre-social' body existing beyond discourse and a socially constructed body, the body as the product of power/knowledge. Biology needs to be welcomed back into the debate without implying biological reductionism or genetic determinism.

One of the figures who has tried to move beyond the mind/body dualism is Maurice Merleau-Ponty. He replaces Descartes' 'cogito ergo sum' with a sentient body-subject. He shows that perception is not just a matter of perceiving the object world because how you perceive depends on where your body is and how it touches and feels. Perception is thus bodily and material.

It is always situated in a physical context and it is always 'of something'. Similarly, emotions such as anger are not just inner states; they involve anger 'about something' and they involve practices such as shaking a fist. So according to Merleau-Ponty, mind/soul and body, subject and world are not split; they are enacted together in every instance of existence.

Erving Goffman is another writer who, it can be argued, breaks away from a simple society/body dualism. In his focus on the micro-practices of daily life he explores how actors deal with walking through spaces, feeling embarrassed, behaving differently in 'front' and 'back' areas. In all these ways, social life is a matter of practical daily competence in which the body plays a central role. Also concerned with the micro-practices of daily life is Pierre Bourdieu. In his ideas of habitus and bodily hexis he examines how the dispositions and postures of the body inculcate social structure into being and existence. For him, the simple injunction 'stand up straight' can embody a whole philosophy of life. But in the end, his is a view that emphasises the socialisation of the body within objective structures. The focus is all on the social determination of the body.

Psychoanalysis is an approach to the body that might be thought again to be dualistic, putting emphasis on natural instincts and drives, and hence tending towards biological reductionism. However, Elizabeth Grosz argues that, for Freud, drives have representational dimensions that both transform and transcend biological instincts. Thus the body is 'rewritten' or 'traced over' by desire. The unconscious, too, is not composed of raw biological instincts but of mental representations that we attach to instincts.

This suggests that Freud can be read as transcending the mind/body duality. The idea of the body as inscribed by culture is found in the work of Jacques Lacan, taking a post-structuralist approach to psychoanalysis. For Lacan, the body is constructed through language and external relationships. While these ideas have been liberating in that they have undermined essentialist accounts of female and male sexuality, the emphasis on language threatens to lurch back into a biology/culture dichotomy in which the second term is privileged, through language. Gilles Deleuze and Felix Guattari take a radical approach in their attempt to dissolve mind/body dualities. Their emphasis on a continual process of becoming, their concern to break down the objects of binary thought into microprocesses, and their focus on contingency and fluidity may all seem attractive, but it can be argued that their account remains dualistic while also dissolving or making disappear the body itself.

Perhaps it is feminism that has been most successful in getting away from the dualism and moving towards an integrated theory of embodiment. Feminism has worked to link lived experience to cultural representation. Much early feminist writing on the body, however, opposed sex (biological) to gender (cultural), and in so doing continued to work within a dualistic framework. More recent, Third Wave, feminists such as Luce Irigaray and Judith Butler, have tried to undermine the dichotomy in favour of a radical approach to 'difference', plurality and fluidity. Irigaray contrasts the singleness, the oneness, of male sexuality with the multiple, diverse and more subtle localities of desire on the female body. Women, she says, need to explore and find a discourse for this plurality and flow. In this she is trying to include both the materiality of female bodily experience and its cultural construction.

Butler focuses on the construction of sexuality within discourse. 'Sex' is an 'ideal construct', and in the performance of sexuality, in materialising sex, regulatory norms are reiterated. Thus the materialising of the body is an effect of power. So 'sex' is not just something I 'have', but it is one of the regulatory norms which makes 'me' possible at all. So the material performance of sexuality reproduces discursive and regulatory powers. Excluded from regulation are those forms of sexuality dubbed by society as 'queer'; and yet queerness can be performed and publically asserted in order to resignify, defiantly, homosexuality and other sexual preferences. Here there is fluidity and contestation – even 'matter' is seen as an effect of power. But once again discourse seems to dominate over bodily experience such as that of pain. In the desire to move away from a 'pure body', perhaps too much is given to discourses of power.

A further, and perhaps more successful, feminist attempt to deal with the relationship between biology and culture with regard to sexed bodies is provided by Grosz. She compares the inscribing of society, culture and language onto the biological body to the writing of words on paper. She says that the model of writing onto a blank slate or smooth white page is inaccurate. As any calligrapher knows, the text that is written depends on the quality and type of paper used. We need to get away from the model of writing on a blank page, to a model of etching, where account is taken of the specifics of the material being inscribed and their effects on the text produced. Body and mind, sexed body and cultural gender construction, all need to be understood as dialectical relations in a process of continual transformation. This is the move from either/or to both/and discussed by Latour in Chapter 7.

## Themes connecting the chapters

The briefly summarised intellectual debate above concerning the mind/body duality resonates with wider public tension between two contemporary views of the body – as subject and as object. This tension is the underlying reason the body fascinates us and has become the centre of so much conflict.

The body as subject is graphically opened up by the body as victim. The naked body, the harmed body appears as vulnerable and in need of our help. The body of the Ice Man discussed by Spindler in Chapter 8 was initially thought to be naked, and this apparent nakedness in the freezing Alpine wastes was an important part of his fascination. The body as subject is the body as self, being, embodied experience. This is Canter's body as person (see Chapter 4). Canter focuses on how we develop ideas about self, personhood, and separate identity.

On the other hand, there is the body as object. In Chapter 8, Spindler discusses how the Ice Man raised questions of ownership – was the body Italian or Austrian. In Chapter 4, Canter describes how some violations of the body involve seeing it as 'object', without personhood. The idea of consumerised bodies draws our attention to the ways that body, body parts and body treatments can become part of a market. This is the body no longer as self, as subject, but as something that can be cut up, bought and sold, transformed, objectified. The micro-investigation of the body within the disciplinary practices described by Foucault lead us to the forensic work discussed by Laqueur and Spindler in Chapters 5 and 8. In all these cases, the body is taken apart and dissected, treated as an object. Indeed the process of dissection has been taken to its logical conclusions in the studies of developmental molecular processes described by Twyman (Chapter 1), leading us to a molecular understanding of how a body is made and organised from simple principles. The publication of other genome sequences in addition to the human genome (see Goodfellow, Chapter 2) demonstrates how alike some of the organising molecules are and hence how universal the organising principles happen to be.

In dealing with the tension between these two views of the body, we have become fascinated with stretching definitions of the body. In many areas of society and culture today, there is a focus on the boundaries of bodies, breaking down the oppositions between bodies and machines, between human and animal bodies, between male and female, between alive and dead bodies, between real and virtual, between whole and dispersed, fragmented bodies, etc. As Pollock shows in Chapter 6, the boundary between art and pornography is now being redefined.

So there is uncertainty and tension about the body in Western society today. But there is also something more far-reaching. The objectification and analysis of bodies raise the possibility of control. In Chapter 2, Goodfellow envisions a future in which there is an enormous ability to define the genetic character of individuals and to predict responses to disease and drugs, etc. All this control leads, he thinks, to a better life. But it also leads to the possibility of genetic control of offspring and genetic populations. One great image of bodies in the twenty-first century is of mass misuse of bodily control. However, a more likely scenario, given some the arguments rehearsed above, is that the modern market and consumerism may also come to define what is alterable via genetic intervention, either at the level of pre-implantation embryo selection or later gene therapy. After all, the same arguments were raised upon the advent of in vitro fertilisation (IVF) and no such organised societal misuse of this process has occurred over two decades of, now very common, use. None the less, the publishing of the Human Genome Project marks a fundamental point in our biological understanding of what defines a body, albeit in an extreme degree, which opens up enormous possibilities for medical investigation and intervention.

So in relation to all this objectification and control there is the return to the body as subject. This is seen in Pollock's shift (Chapter 6) from the female nude as the object of voyeuristic male gaze to the female nude as subject – as an exploration of femaleness in particular social contexts. It is also seen in Canter's example (Chapter 4) of how in medical surgery new drugs and keyhole methods have allowed a greater consideration of the patient as person. It is seen in Laqueur's work (Chapter 5) on the rise of human rights issues surrounding the body. Ethics – who has the right to control or own bodies – is the focus of Chapter 3, by Warnock. Does the Ice Man discussed by Spindler in Chapter 8 have rights over his own body? As Latour (Chapter 7) asks, if the body can be endlessly, and expensively, manipulated, who has the right to have this or that body? What is the moral and democratic perspective we should have on the dispersal and recombination of body parts? What will be an ethical politics of the body?

One view is that a rational consensus can be reached by society to deal with the discoveries of science. This is Warnock's view. Latour in a different way argues that social, emotional, 'secondary' factors can push genetic bodily 'primary' science in certain directions in order to achieve a democratic, negotiated result. Certainly, the body will remain a site of contestation between conflicting claims on the body.

### Conclusion

So, in the end, body and society are not separable. As Warnock says in Chapter 3, how we understand what is meant by the body is social and historical; it is continually changing. She shows that as society and technology change so does the definition of the body.

Pollock in Chapter 6 shows that the boundary between art and pornography has continually changed in law since the mid nineteenth century. Canter (Chapter 4) refers to changes in the way the body has been treated in punishment in recent centuries. He shows how brutal torture of the body, once an acceptable spectacle, is now rejected by many countries.

But it is not enough simply to have an intellectual debate about, as Latour describes it, the dualism between culture and nature, phenomenological and physiological, culture and biology, spiritual and material, subjective and objective. All the authors in this book argue that in the end the need is for a debate about the ethics and the politics of the body. We hope that this volume has contributed to that debate.

FURTHER READING

Clark, W. and Grunstein, M., *Are we Hardwired? The Roles of Genes in Human Behaviour*, Oxford: Oxford University Press, 2000.

Foucault, M., *The History of Sexuality*, vol. 1 *An Introduction*, London: Allen Lane/Penguin, 1979.

Giddens, A., *The Transformation of Intimacy: Love, Sexuality and Eroticism in Modern Societies*, Cambridge: Polity Press, 1992.

Grosz, E., *Volatile Bodies: Toward a Corporeal Feminism*, Indianapolis: Indiana University Press, 1994.

Ridley, M., *Genome: The Autobiography of Species in 23 Chapters*, New York: HarperCollins, 2000.

Turner, B., *The Body and Society*, London: Sage, 1996.

Williams, S. and Bendelow, G., *The Lived Body*, London: Routledge, 1998.

Wilmut, I., Campbell, K. and Tudge, C., *The Second Creation: Dolly and the Age of Biological Control*, New York: Farrar, Straus and Giroux, 2000.

# 1  Building the Body – The Molecular Basis of Development*

RICHARD M. TWYMAN

## Introduction

One Saturday afternoon in the summer of 1997, I answered a knock at my door to find a Jehova's Witness standing under the porch. This was not the first time a Witness had called at my house, and I am sure it will not be the last. My usual reaction to these people, if I have time, is to engage them in polite conversation. I enjoy pitting my unshakeable belief in science against their unshakeable belief in God, and I am genuinely interested in their absolute faith, which makes them willing to give up their time to calling at strangers' houses. On this occasion, I invited the caller in for coffee and we spent the afternoon chatting about our different beliefs. After a while, he said to me that he could never trust science because it could not explain why a cell in your right hand was different from a cell in your left hand, yet God had made it so. Perhaps he was right, but as a teacher of developmental biology to undergraduates in Cambridge, I felt it was my duty to explain at least some of the principles of limb development that scientists have deciphered over the past twenty years. In the end, I promised to read his Bible if he promised to read a text book called *Developmental Biology* by Scott Gilbert, which is almost as big as the Bible, but has more pictures and references. To give the man his due he took the book and returned it a week later, with the comment that he had found it 'pretty hard going' but enlightening. And yes, I also fulfilled my side of the bargain. Developmental biology aims to explain the fundamental cellular and molecular basis of how a living body is constructed, starting with a single cell, the fertilised egg. This process involves growth, the specialization of cells into different types, cells

* The first lecture in this series was given by Christiane Nusslein-Volhard.

12

organizing themselves into patterns and forming defined structures. It has fascinated and mystified human beings since the dawn of history. However, over the past thirty years or so, many of the fundamental processes involved in development have begun to unravel under the relentless attack of experimental investigation. Some of the remarkable processes that occur during development, and their molecular underpinnings, are discussed in this chapter.

## Building the Round Church

Peter Goodfellow (Chapter 2) has likened the body to a large and complex building, where different types of bricks are arranged in patterns rather in the same way that different cell types are organized in the body. This analogy breaks down when we consider how bodies and buildings are built. The Round Church in Cambridge, and all other buildings, were (and continue to be) constructed brick by brick from the foundations upwards. Conversely, a body is not made starting with the cells of the feet and building towards the head. Rather, growth and development occur gradually in all areas simultaneously. The simple reason for this is that the body is alive even as an embryo and must be functional. Conversely, until it is finished, it is quite acceptable for a building to lack a roof or windows, since it is not used until construction is complete.

The miracle of development was appreciated as long ago as the fifth century BC, when Greek philosophers such as Hippocrates, and later Aristotle, debated its mysteries. Until the eighteenth century, scientists argued over two major theories – pre-formation and epigenesis. In the theory of pre-formation, the human body was thought to be ready formed in either the sperm or the egg. Pre-formation was a comfortable theory because it required no explanation for the exquisite structure of the body other than the hand of God. All the organs and systems of the adult body were present in miniature at conception, and development, stimulated by the meeting of sperm and egg, involved nothing more complex than growth. Some scientists at the time even claimed to be able to see tiny human bodies curled up in the head of the sperm! At the time, the existence of cells and atoms was unknown and there was no concept of a 'minimum size'. Each generation of human beings was considered to be pre-formed in ever decreasing dimensions within the undeveloped sperm and eggs of the embryo. It was thought that the whole of humankind had been created at one point in time, like a series of Russian dolls (Figure 1), beginning with Adam himself.

Despite its attractive simplicity, the pre-formation theory provided no account for certain observations, such as the intermediate skin colours of

**Preformation**

**Epigenesis**

FIGURE 1. Pre-formation versus epigenesis. In pre-formation, the detail of each individual is ready formed and generations are packaged like a series of Russian dolls. In epigenesis, the detailed pattern of the body forms gradually, like a photograph gradually increasing in resolution.

children produced by mating between blacks and whites. If embryos were pre-formed, there should be no mixing of characters in this way. The theory was eventually discarded when it was discovered that living organisms were made of cells, and that development started with a single cell that went on to divide many times. In the light of such revelations, it was no longer possible that entire individuals could be pre-formed within the eggs of the eggs of the eggs of an embryo for countless generations. The alternative theory of epigenesis suggested that development involved a progressive diversification of cells and structures concomitant with an increase in cell number, to form all the organs and systems of the body anew in each generation. The embryo is initially very simple, a few cell types organized in a crude pattern. As development proceeds, more cells are formed and the pattern gradually becomes more finely detailed. This theory fitted the data perfectly, but raised a number of important questions that have begun to be addressed only in the last few decades.

> How do cells become different from each other during development?
> How do cells become organized into patterns to form individuals with a similar general appearance?
> How do cells form particular structures and shapes in developing tissues?

### Differentiation – diversifying the body

The human body begins life as a single cell, a fertilized egg. However, the adult human contains many hundreds of specialized cell types, for example neurons in the nervous system, myotubes in the muscles, adipocytes to store fat, erythrocytes to carry oxygen in the blood, keratinocytes to form the outer layer of the skin, hepatocytes in the liver and osteocytes to produce bone. All of these cell types arise from the same egg. At the beginning of development the egg divides a number of times in rapid succession to form a ball of similar cells, a process termed cleavage. After this stage, the cells begin to undergo a gradual process called differentiation, where they diversify into different cell types. Just after cleavage, the human embryo diversifies into only three cell types, called ectoderm, mesoderm and endoderm. These diversify further in a hierarchical manner, eventually producing the full spectrum of cell types in the adult.

**How** is differentiation achieved? The **important** point to grasp here **is** that specialized cells are different from **each** other because of the repertoire of proteins they contain. Proteins are macromolecules composed of amino acids. The sequence of amino acids determines the chemical properties of the protein and thus how it functions in the cell. For example, erythrocytes can carry oxygen through the blood because they produce large amounts of the protein haemoglobin. Other cells do not produce this protein, and therefore cannot carry oxygen. Muscle cells owe their extraordinary contractile ability to proteins such as skeletal actin and myosin. Other cells do not produce these proteins and therefore cannot carry out the specialized functions performed by **muscle** cells. In turn, proteins are encoded by genes and with few exceptions every cell **in** the body contains **the** same genes, which are also the same genes as are found in the fertilized egg. The principle of differentiation involves controlling how those genes are used to make proteins in different cells of the **body**. Muscle cells and immature red blood cells contain the same globin genes that make up haemoglobin, but the genes are switched on only in the developing red blood cells. Similarly, the genes encoding the contractile proteins are switched on only in muscle cells. This selective use of information enables the same genome (the full complement of DNA possessed by a given organism) to produce cells with very different functions. Go back and read this paragraph again, but this time read only the bold words. The selection of particular words brings out a sentence with a unique meaning: **how important is each muscle in the body**? The selective use of genetic information during development can similarly bring out a number of unique cell types.

Plate I shows a section through the developing spinal cord of a mouse embryo. The structure in the centre is a dorsal root ganglion, a conglomeration of sensory neurons that receive inputs from sensory organs around the body and feed their axons into the central nervous system. At this stage of development, certain genes are just beginning to be switched on as part of neuronal differentiation, and the corresponding proteins are beginning to be made. Antibodies raised against such proteins can be joined to a fluorescent label and this can show specifically in which areas of the embryo the proteins are made. In this particular example, the protein is called neuron-specific enolase. This is expressed at high levels in neurons relatively late in their differentiation. All over the embryo, and throughout development, different genes are being switched on in different cells to make them structurally and functionally distinct.

## Pattern formation – organizing the body

The Round Church in Cambridge is constructed from different types of brick, and the body is similarly made of different cells, but different cells are not enough. In the Round Church, different types of brick are found in specific places, such as around windows and doors, and similarly, in the body, particular cell types are restricted to certain locations. For example, the eye contains various specialized cell types that are found nowhere else in the body, such as retinal ganglion cells, rods and cones. Therefore, development cannot rely solely on the creation of different types of cells – they must also be organized in a distinct pattern. Both differentiation and pattern formation are essential for development. It is all very well having rods and cones and retinal ganglion cells, but if these are dispersed randomly throughout the body, there would be no eyes and no-one would be able to see! For correct function, these cell types must be localized in the head and organized in a specific manner to form the eye.

Pattern formation in development has fascinated scientists and lay people alike, for when the process goes wrong it generates the most bizarre and unusual effects. In humans, this ranges from mild conditions such as polydactyly, where individuals may have six, seven or even more fingers on each hand, to severe defects such as cyclopia (having a single eye), which in previous centuries used to provide the major attractions in barbaric and exploitative freak shows. Scientists work on organisms that are simpler than humans to establish the principles of development, and the species that has provided the most information concerning the process of pattern formation is the fruitfly *Drosophila melanogaster*. Indeed, I think the thing that most attracted me to developmental biology when I finished my A-level examinations was the *Drosophila* mutant called *Antennapedia* pictured on the genetics course leaflet for Newcastle University. As its name suggests, this unfortunate insect has legs growing out of its head in place of antennae. The cells in the antennae-forming region of the fly larva somehow 'think' that they belong further down the body, and so develop as though they were sticking out of the thorax rather than the head.

Pattern formation is all about telling a cell where it is in the embryo in relation to other cells, so that it can behave in the appropriate manner for its position and form the correct structure. If you are lost in a strange city, you would consult a map to find your whereabouts, and the map is based on a framework of coordinates, in turn based on the compass points north, east, south and west. Similarly, the cells in the embryo are organized in relation to a framework of coordinates based on the principal axes of the body. The craniocaudal axis runs from head to tail, the dorsoventral axis from back to belly, and a third axis runs from left to right. If a cell is 'aware' of its position along all three axes, it can be unambiguously assigned a location in the embryo and can behave in accordance with that position to generate the correct regional structure. The first aspect of pattern formation in development is therefore the establishment of these axes. Remember that the egg is basically a symmetrical cell, so axis specification must involve some symmetry-breaking process. A wide variety of different mechanisms is used. In *Drosophila*, and many other invertebrates, the egg develops surrounded by maternal cells and these place molecules inside the egg that become asymmetrically distributed to define the head and tail ends, and the dorsal and ventral surfaces. In amphibians, the place where the sperm enters the egg plays a major role in determining both the craniocaudal and dorsoventral axes. In chickens, the tail end of the embryo is determined by gravity in a critical two hour period as the egg rotates on its way down the oviduct. In mammals, the dorsoventral axis of the embryo is thought to be defined by the orientation of the embryo with respect to the cells that will form extraembryonic membranes. The left–right axis in mammals is the most intriguing of all, since it is established after the other axes have formed by the beating of tiny hairs in a central region of the embryo called the node, moving tiny amounts of fluid to the left-hand side of the embryo and activating different genes on different sides.

Once the axes are set up, cells need to be made aware of their position along the axis (Figure 2). This is often achieved by cells at one end of the axis producing a certain molecule that forms a gradient, so that high concentrations are formed at one end of the axis and low concentrations at the other, and all possible intermediate values in between. The surface membranes of animal cells bristle with receptors for various molecules, allowing them to react to proteins and other molecules in the environment. This, for example, is how cells respond to hormones. The receptors are usually linked to internal signalling pathways that eventually result in particular genes being switched on. Cells may switch on certain genes only at high doses of a par-

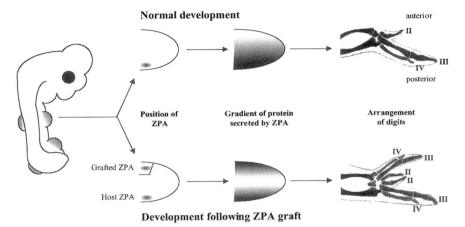

**Normal development**

anterior

II

III

IV

posterior

Position of
ZPA

Gradient of protein
secreted by ZPA

Arrangement
of digits

Grafted ZPA

Host ZPA

IV III

II
II

III

IV

**Development following ZPA graft**

FIGURE 2. An experiment showing how cells get to know their position in a
developing embryo. A chicken wing has three digits, conventionally
termed II, III and IV, with II anterior and IV posterior. Normal development involves a signalling centre called the zone of polarizing activity (ZPA) that specifies the posterior of the limb. A protein
synthesized here forms a gradient, and cells respond to this gradient
at different concentrations by forming different digits (at high concentrations, digit IV is formed, at low concentrations, digit II is
formed). If a ZPA is grafted from one limb bud to the normal anterior side of a host limb bud that already has its own ZPA, a double
gradient is established, and cells that would normally form anterior
digits are respecified to form posterior ones. The limb that develops
has double the normal number of digits arranged in a symmetrical
pattern.

ticular activator and other genes at low doses. Therefore, cells at different
positions in a concentration gradient of an activating molecule may begin to
produce different proteins and this is what causes them to behave differently according to their position. By artificially manipulating such gradients,
scientists can reorganize the pattern of developing structures such as limbs,
for example reversing the orientation of the digits and generating extra
fingers or toes (Figure 2).

Cells in different positions along an axis switch on different combinations
of genes that guide cell behaviour, causing them to generate structures appropriate for their position. If these genes are disrupted by mutation and the
cell cannot switch them on, it will be forced into the behaviour appropriate
for a different region of the body, resulting in the development of a misplaced

structure, such as legs in place of antennae. This is known as a homeotic mutation, a mutation that causes one body part to develop with the likeness of another (Figure 3).

In one of the most exciting discoveries in the history of developmental biology, the genes responsible for homeotic mutations in *Drosophila* were found to be grouped closely together in one cluster and encode proteins that control how other genes are switched on and off. The genes were expressed in overlapping patterns so that each particular cell could be given a unique 'positional value' based on the number of genes that were expressed. The simplest way to explain this is to liken the system to a binary code, in which each expressed gene is represented by 1 and each silent gene is represented by 0. In this way, individual cells can be given different binary codes according to their position along the axis, and this will enable them to behave in the appropriate manner to generate the correct regional structure (see Plate II). Mutations in the genes would change the code, so the cell would think it was in an entirely different position, and a different body part would develop. Even more exciting was the discovery that all animals possess a similar set or sets of genes that control cells' positional values. In humans and most

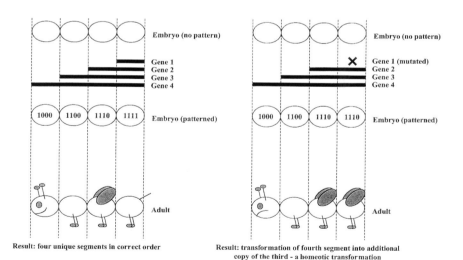

FIGURE 3. Patterning in a hypothetical insect: how cells know where they are and what to do. Combinations of genes are expressed in overlapping patterns to give each segment of the insect a unique code (shown as a binary code (ones and zeros)), which controls how the cells behave to make appropriate structures (left panel). If these genes are mutated (right panel), the code is changed and cells behave inappropriately for their position, generating body structures in the wrong place.

other vertebrates, there are four distinct sets of *Hox* genes, but the characteristic overlapping expression patterns are similar to those seen in flies, and mutations in the genes also affect the position of different body structures. For example, disrupting certain *Hox* genes in mouse embryos causes the cells of the lumbar vertebrae to believe they are actually part of the thorax, resulting in mice with an extra set of ribs! The remarkable conservation of the *Hox* genes among animals suggests that all animals, however diverse, are simply variations on a similar developmental theme.

### Morphogenesis – structuring the body

Morphogenesis means the creation of form and this can be regarded in many ways as the last piece in the developmental puzzle. We have discussed above how a developing cell decides what sort of cell it will become in the adult, and how it finds out where it is in relation to other cells. Now, endowed with this information, the cell goes on to behave in a manner that generates the particular structure appropriate for that part of the body. The way in which cells behave to form particular structures is termed morphogenesis.

The way in which cells form different shapes is quite remarkable. Plates II and III show the structures formed by cells in the developing fruit fly and mouse brains, respectively. In both cases, there is an exquisite architecture. Cells simply dividing and dividing to produce more cells would form an amorphous blob, so the behaviour of cells must be precisely controlled to allow such beautiful yet functional organs and tissues to form. There is a wide variety of morphogenetic mechanisms used during development, some of which are listed below:

Cell division – cells can divide at different rates in different parts of the embryo, and can divide in different planes to generate particular structures. The plane of cell division is particularly important in plants, where other types of cell movement are restricted by the cell walls.

Changes in cell size and shape – this can cause the folding and buckling of cell sheets to generate curves and hollows.

Cell fusion – cells fuse together, for example, during muscle development.

Cell adhesion – the way in which cells stick together, through adhesion molecules displayed on their surfaces, plays a predominant role in maintaining tissues and tissue boundaries, and causing cells to move in relation to each other. The loss of cell adhesion causes cells to disperse.

The extracellular matrix – this is a network of molecules secreted by cells into their immediate environment. Interactions between cells and this substrate can maintain cell sheets and provide a surface over which cells can migrate.

Cell death – surprisingly, cell death also plays an important role in development, as discussed below.

Cell behaviour can therefore take many forms, and as an example, we return to Peter Goodfellow's original analogy of the Round Church (Chapter 2). The church is made of bricks stuck together with cement. Similarly, a body is made of cells that have special molecules on their surfaces that enable them to stick together, producing tough and durable tissues. While most bricks are rectangular, occasionally there are bricks that are wider at one end than the other, and these are used to make specialized structures such as arches. Similarly, by changing the shape of a cell, the tissue can be made to bend and distort to form a tube. In exactly this way, many of the tubular structures of the body are formed, for example, the neural tube that gives rise to the brain and spinal cord.

A house without windows and doors is not much use, and similarly there is as much developmental potential in leaving gaps in the body as there is in filling it with cells. In the case of the house, the bricks are simply left out of the window frames. However, since development begins with a single cell, it is not possible to leave gaps where gaps are required in the adult. Gaps must be formed by controlling cell behaviour. A good example of this process is the formation of the gaps between your fingers. The hand, indeed the whole arm, begins development as a small protruberance from the side of the embryo's body known as the limb bud. As development proceeds, the limb bud extends and begins to form the structures of the mature limb. The hand itself is initially a circular pad with no fingers. Between the fourth and eighth weeks of gestation, cells in the interdigital regions are instructed to die as part of the overall developmental programme. These cells are in essence no different from the cells that eventually form the fingers, but because of their position in the hand their role in development is to die rather than form the bone, muscle and skin of the finger. Cell death is an important feature of development generally, particularly in the nervous system, where over half of the neurons originally created during development are killed off in the process of establishing and refining the complex circuitry required for a functioning nervous system. Therefore, despite the claims made by Captain

James T. Kirk of the Starship *Enterprise*, in development at least, the needs of the few do not outweigh the needs of the many.

## The future

In the past, individual genes with important roles in development were identified on the basis of their mutant phenotypes, i.e. the effect on the organism when the gene was mutated. Scientists have slowly pieced together the immense puzzle of development using the fragments of information generated by the analysis of individual genes in various model organisms, including the fruitfly *Drosophila*, a nematode worm called *Caernorhabditis elegans*, amphibians such as the frog *Xenopus laevis*, the chicken, the mouse and most recently, a highly versatile zebrafish, *Danio rerio*. The puzzle is not yet complete, but the future for developmental biology looks bright, since genome projects are underway for a number of these model organisms, and in the case of *C. elegans*, the genome sequence is complete. I believe that, one day, it will be possible to describe in detail the entire process of development, from egg to adult, in terms of genes, proteins and cell–cell interactions, and perhaps manipulate developmental processes for our own ends. This will make it possible to put right birth defects, regenerate amputated limbs, replace burned skin, and cure diseases of ageing, such as Parkinson's disease and Huntington's disease. The molecular biologist Edwin Chargraff once expressed his concerns that tinkering with biological processes was akin to a child testing a toy to destruction. However, by learning how a toy is built, it will be possible not only to repair broken toys but to make the toys better and more resilient in the first place.

FURTHER READING

Gould, S. J., *Hen's Teeth and Horse's Toes*, Middlesex: Penguin Books, 1983. [Two chapters in this book provide very entertaining and accessible discussions on development: Chapter 14 (Hen's teeth and horse's toes) and Chapter 15 (Helpful monsters).]

Lawrence, P. A., *The Making of a Fly*, Oxford: Blackwell Scientific Publications, 1992. [An excellent book delving a little deeper into *Drosophila* development; much used by students.]

Nusslein-Volhard, C., 'Gradients that organize embryo development', *Scientific American* **275** (1996) 54–55. [A nice account of pattern formation in *Drosophila*.]

Slack, J. M. W., *From Egg to Embryo*. Cambridge: Cambridge University Press, 1991. [A comprehensive text for those seeking a little more information.]

Twyman, R. M., *Instant Notes Developmental Biology*, Oxford: BIOS Scientific Publishers, 2000. [A brief summary of major research areas in developmental biology, with a guide to current literature on the subject.]

Wolpert, L., *The Triumph of the Embryo*, Oxford: Oxford University Press, 1991. [This book provides a wealth of accessible information on the importance of developmental biology and how it is studied.]

## 2  Mapping the Body – the Human Genome Project

PETER N. GOODFELLOW

### Introduction

Historians will soon be arguing about the most important achievements of the twenty-first century. Some will claim that quantum physics should have pride of place. Quantum physics has positioned us in the universe and provided society with the moral challenge of the fusion bomb. Others will argue the importance of placing a man on the moon. Still others will argue for the invention and evolution of the computer. I will not argue against the contributions of Einstein, Bohr (my favourite because he had great feet as well as a great mind; he was awarded the Nobel Prize and played football for Denmark), Rutherford and the other physicist-giants, nor will I dispute the legacies of von Braun (a mixed legacy if you come from London), Gagarin, Shepard, Turing and Gates. But I will contend that the most important achievement of the twentieth century was the discovery of the structure of the chemical deoxyribonucleic acid. The name is such a mouthful that deoxyribonucleic acid is universally known as DNA. I will also make the claim that the most important discovery of the twenty-first century will be made in the first few years of the new millennium.

To state my case, I will need to provide an introduction to biology, describe the structure of DNA and give examples of how knowledge of the sequence of DNA will impact on society.

### The Round Church

The analogy between the body and a building pervades religious imagery and the body is often portrayed as the 'temple of the soul'. One of my favourite ecclesiastical buildings is the Round Church in Cambridge. This

25

circular building is over 800 years old and is a wonderful example of the fusion of art and science. Like most buildings, it is made of smaller units called bricks (see Figure 1). Careful inspection reveals that not all bricks are the same. The walls are made of one type of brick and other parts are made of other types of brick, for example the bricks around the door and windows are distinct. Even a casual look suggests ten or twenty types of brick are present. In a similar way, all living things are made of bricks but the 'bricks'

FIGURE 1. The Round Church, Cambridge. This magnificent building is made of bricks – many different types of brick.

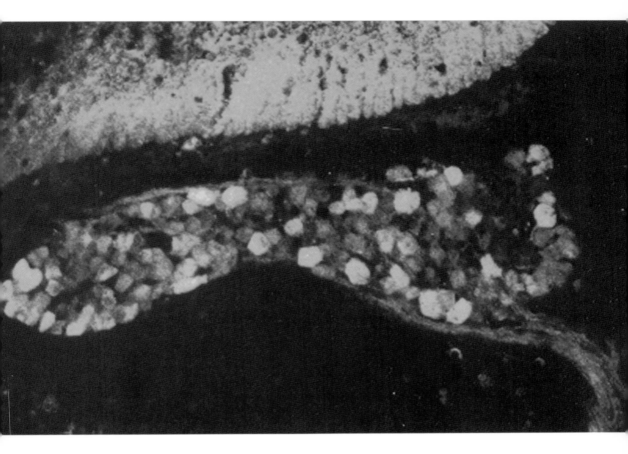

Staining for the neuron-specific enolase protein in the dorsal root ganglion of a mouse embryo (14 days' gestation).

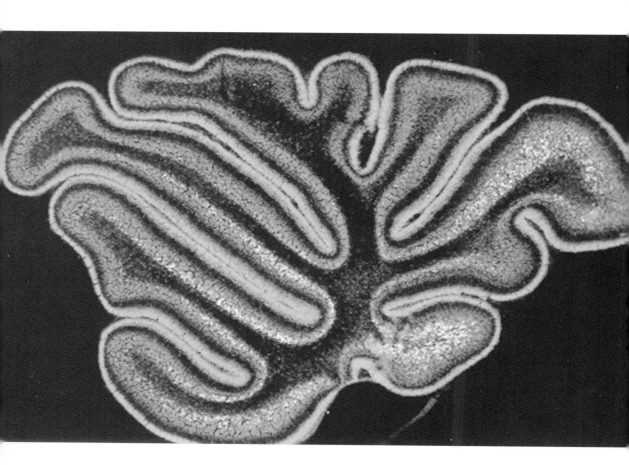

NSE staining of dorsal root ganglion (E14 mouse embryo).

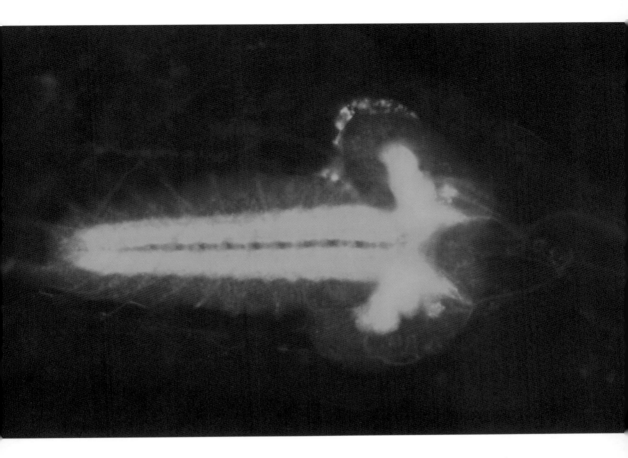

Green fluorescent protein staining of the central nervous system in *Drosophila* larva.

The head and shoulders of the human mummy emerging from the
glacial ice as discovered towards 1.30 p.m. on 19 September 1991.

are called cells. In simple organisms – unicellular organisms – the whole animal or plant may comprise a single cell. In multicellular animals, like humans, there are many different types of cell. There are nerve cells in the brain, skin cells in skin, muscle cells in muscle, fat cells in fat and so on. Altogether it has been estimated that there are about 100 million, million cells in a human body and these can be classified into at least 10 000 distinct types. The master builder constructing the Round Church had to communicate with the common labourers about where to place which bricks. Today an architect controls this process by drawing up a blueprint. This is where the analogy between a building and the human body breaks down. Each cell in the body contains its own copy of the complete set of instructions for making a complete creature. This set of instructions is present within a specialised structure called the nucleus, which is located within each and every cell. Ian Wilmot and colleagues achieved startling proof of this statement when they transferred the nucleus from a cell from an adult sheep to a sheep's egg from which the original nucleus had been removed. This is how Dolly the sheep was created, by cloning. The other difference between cells and bricks is that cells can replicate and produce new cells. Sometimes the new cells are replicas of the parent cell and sometimes the cells produced differ from those of the parent. All the cells and all the cell types in a human derive ultimately from the single fertilised egg cell.

Just as the building is made of bricks and the body is made of cells, the cell is made of units called proteins. In turn proteins are made of amino acids. There are twenty different types of amino acid found in proteins and the linear order of these amino acids is specific for each protein. A typical cell is made up of about 10 000 different proteins. Proteins act as structural components of cells and as catalysts of chemical reactions.

In the first part of the twentieth century there were furious debates about the nature of instructions used to build an organism. Many believed that only proteins had the complexity needed to code the instructions, while others argued the claims of carbohydrates and nucleic acids. In 1944, Avery, Macleod and McCarthy ended the argument in a very persuasive way – they did an experiment. By adding to benign bacteria purified DNA taken from a disease-causing bacteria, they were able to convert the bacteria into the disease-causing form. DNA could encode biological information but how?

**'It must be late in the day for such pygmies to cast such long shadows'**

In 1953, a physicist, Francis Crick, working with a biologist, James Watson, solved a chemical problem. They solved the structure of DNA (Figure 2). As in all good stories there was drama, conflict and many other characters, but the inspirational leap to divine the structure belonged to these two. Not everyone was generous in praise of the two young men; Chargaff's unkind quip must rank amongst the unkindest and erroneous remarks ever made. In the true traditions of Cambridge University, the building where Crick and Watson worked has been turned into a bicycle shed – albeit a bicycle shed with a plaque.

Naturally occurring nucleic acids are composed of units called bases; the four bases that make up DNA are represented by the letters A, T, C and G. Watson and Crick realised that DNA was made of two chains of units

FIGURE 2. Watson and Crick and their model of DNA. James Watson (on the left) and Francis Crick with the famous model. As a student I handled a piece of the model – I was greatly encouraged to discover that there was a corrected error in the placement of nitrogen and carbon atoms in the purine rings.

arranged in an anti-parallel mirror image of themselves. Every time A appeared on one strand it was paired with T on the other strand (a base-pair), every time C appeared on one strand it was paired with G on the other. This structure and pleasing symmetry immediately offered solutions to the problem of encoding information and copying the information. In the replication of DNA, each strand could be used as a template for constructing a new molecule and the order of the bases could be a code. Eventually it was shown that the code was read in groups of three, which specified the order of amino acids in proteins. One unit of information was equated with the gene and specified one protein. The code is not read directly but is first transcribed into a messenger molecule, a nucleic acid called mRNA (messenger ribonucleic acid). The messenger is translated to produce proteins that make up the cell. The description of biology in the form 'DNA is transcribed into RNA, which is translated into protein' is known as the Central Dogma.

One of the glories of DNA is that the two chains are arranged as a helix, hence the names double helix and golden helix, which are often employed by the media. As a teenager in Cambridge, I once went on a blind date. Rowena had arranged this for me. I was in love with Rowena but she was in love with her horse. Rowena introduced me to Gabrielle with the hope that it might leave her more time for the horse. We met in a coffee bar. Gabrielle was both comely and kind; she asked kindly what I did. It seemed obvious to me that I was a schoolboy but I recognised that sophistication was needed. 'I am a biologist', I said. 'My father is a biologist; he discovered the structure of DNA', she rejoined. The blankness of my face prompted her to add 'You know – the Golden Helix'. 'Oh, yeah, that DNA', I answered. If only I had known, perhaps I could have become Francis Crick's son-in-law.

### The genome and sailing

The complete set of instructions for an organism is known as the genome. Genome sizes vary from a few million base-pairs for bacteria to three billion base-pairs for a human and even more for some plants. If it were possible to determine the sequence of a genome it would provide a definition and a description of the organism from which the sequence was derived.

In Cambridge, at the same Institute where Watson and Crick worked, Fred Sanger was inventing a way to determine the sequence of amino acids in proteins. Eventually, he was able to provide a complete sequence of the medically important protein insulin. For this achievement, Sanger received the Nobel Prize for Chemistry in 1958. For most of us, providing the key to a medically important question and earning a Nobel Prize would be achievement in excess of our wildest dreams. Most of us would accept the big 'Prize' and then go sailing. Not Sanger, over the next 22 years he developed techniques for sequencing nucleic acids. This work culminated in a new technology called by the prosaic name of dideoxy chain termination sequencing. For this work Sanger also received a Nobel Prize. In 1989, Sanger retired from science, went home, built a boat in his back garden and then went sailing.

The sequencing method invented by Sanger has not been bettered. New enzymes have been employed; the technique has been converted from radioactive labels to fluorescence labels and the whole process has been automated. Nevertheless, Sanger's method is the method being used to identify the exact order of the three billion base-pairs that make the human genome. A fitting tribute to Sanger has been to name the world's largest sequencing factory – the Sanger Centre (Figure 3).

FIGURE 3. The Sanger Centre at Hinxton in Cambridgeshire, the home of the British contribution to the Human Genome Project. The major sponsor of the research on this site is the Wellcome Trust.

## The Human Genome Project

The human genome comprises about 100 000 genes and three billion base-pairs. The DNA is arranged in structures called chromosomes. The human genome is made up of twenty-two pairs of chromosomes known as autosomes and a pair of sex chromosomes. The total length of DNA in any cell is about two metres; coiling condenses this so that chromosomes are visible with the light microscope during division of cells.

I started my scientific career, seven years after meeting Gabrielle, by mapping human genes to specific chromosomes. I am proud to have assigned the gene encoding beta-2 microglobulin to human chromosome 15, my first contribution to the scientific literature. I was only one of many hundreds of investigators who had been studying the organisation, structure and sequence of the genome for decades. In that sense, the Human Genome Project is not new. The defining moment of the Human Genome Project was when the community of scientists believed that it might be possible to complete the sequencing of the whole genome in the foreseeable future. All that was needed was money, motivation and political support. In retrospect, it was inevitable. A complete draft of the DNA sequence of the human genome was published in the spring of 2001. Completion of the Human Genome Project is likely to be the most important event of this century.

Not only the human genome but the genomes of less complex organisms are being completed. Many bacterial genomes have been finished as well as the genomes of baker's yeast, a small worm and the fruit fly. This information is a test bed for understanding how to read the instructions for a human.

All belief systems constrain our interpretations of the world. Until 1967, in the State of Tennessee, the law stated:

> It shall be unlawful for any teacher in any of the Universities, normals and all other public schools of the state which are supported in part or in part by the public school funds of the state, to teach any theory that denies the story of divine creation of man as taught in the Bible, and to teach instead that man has descended instead from a lower order of animals.

(Butler Act, 1925)

If you accept that the sequence of the genome contains the instructions for making a human, it must constrain and influence how we interpret ourselves.

The sequence of the human genome is finished and is available to every school child who connects their computer to the internet. Here is the complete instruction set for making all the proteins for making all the cells for making a human. We know how to read the code for protein sequences but we do not know yet how to read the code that controls where and when particular proteins are made. How can we exploit this new information? What does it mean for society?

**What are little boys made of? Slugs and snails and puppy dogs' tails.**
**What are little girls made of? Sugar and spice and all things nice.**

The origin of the differences between the sexes has been the subject of debate since the ancients postulated that ardour and position during coitus was the trigger that guaranteed the birth of a male. Even in this century, my mother claims to be able to predict the sex of a child by the position of the fetus *in utero*. My children have been taught a different theory, a theory that is rooted in the sequence of the human genome.

The genome is organised in units called chromosomes. Each of us has 46 chromosomes arranged as pairs, one member of each pair is inherited from each parent. Chromosomes are named according to size from chromosome 1 to chromosome 22 (actually chromosome 21 is smaller than chromosome 22 but that is a different story), except for the last pair, which are unique because they can be dissimilar from each other – these are the sex chromosomes. In females, the sex chromosomes are the same and are known as X chromosomes. Males have an X chromosome and a much smaller Y chromosome. Sex determination is simple: if you inherit a Y chromosome from your father, you will be male; if you inherit an X chromosome from your father you will be female. Both sexes inherit an X chromosome from their mothers. This simple story makes a very strong prediction: the Y chromosome must contain either a gene or many genes responsible for maleness.

As in all simple stories there are complexities. One in every 20 000 men does not have a Y chromosome. At first sight, this is a paradox. Two explanations were proposed: either these XX males had a hidden fragment of the Y chromosome that was sufficient to produce male differentiation or an alteration had occurred in their DNA instructions that caused male differentiation in the absence of a Y chromosome. Unlike in most thrillers, both explanations of the plot turned out to be correct. By studying the genomes of XX males it was possible to find a small fragment of the Y chromosome that was responsible for inducing male differentiation. The surprise was that the fragment was very small. The gene responsible for male sex determination is only 612 base-pairs long and codes for a protein that is only 204 amino acid residues (linked amino acids) in length. The SRY gene, the difference between men and women, is presented below:

ATGCAATCATATGCTTCTGCTATGTTAAGCGTATTCAACAGCGATGATT
ACAGTCCAGCTGTGCAAGAGAATATTCCCGCTCTCCGGAGAAGCTCTT
CCTTCCTTTGCACTGAAAGCTGTAACTCTAAGTATCAGTGTGAAACGG
GAGAAAACAGTAAAGGCAACGTCCAGGATAGAGTGAAGCGACCCATG
AACGCATTCATCGTGTGGTCTCGCGATCAGAGGCGCAAGATGGCTCTA
GAGAATCCCAGAATGCGAAACTCAGAGATCAGCAAGCAGCTGGGATA
CCAGTGGAAAATGCTTACTGAAGCCGAAAAATGGCCATTCTTCCAGGA
GGCACAGAAATTACAGGCCATGCACAGAGAGAAATACCCGAATTATA
AGTATCGACCTCGTCGGAAGGCGAAGATGCTGCCGAAGAATTGCAGTT
TGCTTCCCGCAGATCCCGCTTCGGTACTCTGCAGCGAAGTGCAACTGG
ACAACAGGTTGTACAGGGATGACTGTACGAAAGCCACACACTCAAGA
ATGGAGCACCAGCTAGGCCACTTACCGCCCATCAACGCAGCCAGCTCA
CCGCAGCAACGGGACCGCTACAGCCACTGGACAAAGCTGTAG

(I know that there are 615 bases listed and these should correspond to 205 amino acid residues but the last three bases -TAG- is code for stop)

To prove that this short sequence is the only difference between the sexes, Robin Lovell-Badge and his co-workers took the mouse version of the sequence above and added it to the genomes of fertilised eggs that had two XX chromosomes. The female embryos were converted into males. These male embryos grew up to be male mice.

DNA is a long thin molecule rather like a piece of molecular string. Obviously, as a molecule, it is too small to be seen. If it were magnified so that it had the thickness of a piece of string, the human genome would be

3000 kilometres long and would stretch from Cambridge to Moscow. Using this analogy, it is possible to illustrate how similar are different genomes. The human and the chimpanzee genomes differ by 1% or thirty kilometres – the distance between Cambridge and the City of Ely, on the way to Moscow. Any two sets of human instructions, chosen at random, differ by 0.1% or three kilometres – the distance between the centre of Cambridge and the village of Girton on the outskirts. The sex-determining gene would cover less than a metre or one human step on the journey to Russia. The random genetic differences between two humans are 3000 times greater than the genetic differences between men and women and yet our society imposes a biological destiny on women because of this infinitesimal difference. Cambridge University has never had a female Vice-Chancellor and few are the women who are allowed by right into the Fellow's Room at 6 Carlton Terrace (the home of the Royal Society).

## Variation, disease and the National Health Service

One of the greatest glories of being human is to be able to enjoy the differences between humans. Some of us are fat, some are thin; some have blond hair, some have dark hair, some have red hair, some have no hair; some have black skin, some have brown skin; some are tall, some are short – the variations are endless. This variation is due to two fundamental causes: the precise sequence of our genomes and the environment to which we are exposed. On average, we differ from our neighbours by one base-pair every 1000 base-pairs. Occasionally, these differences will cause differences in either protein expression or protein function and this will lead, in turn, to changes in cells that can lead to differences in appearance. Similarly, changes in environment can lead to changes in physical appearance. If you eat large amounts of food your weight will increase. In real life, things are more complex because genes can interact with genes, environment can interact with environment and genes can interact with environment. Although everyone gains weight if they eat more, some people gain far more weight and become obese eating exactly the same amount as others who gain but a few pounds. The environment is the same but the genes are different. In the same way that appearance is controlled by genes and environment, so is disease. At one extreme, an environmental catastrophe can cause disease, for example, being run over by a bus. At the other extreme, inheritance of a single defective gene can cause devastating disease, for example muscular dystrophy. Most disease, however, is caused by complex interactions between genes and environment.

At the molecular level, disease is expressed through alterations in protein activities. Disease can be due to loss of function, reduction in function, change of function, increase in function and alteration in time and place of expression of proteins. An example of a simple genetic disease is sickle cell anaemia. Haemoglobin is an essential protein that makes blood red and carries oxygen around our bodies. This protein is composed of two chains, alpha and beta globin. The sequence of the beta globin protein is presented below (each type of amino acid is represented by a single-letter code):

MVHLTP**E**EKSAVTALWGKVNVDEVGGEALGRLLVVYPWTQRFFESFGD
LSTPDAVMGNPKVKAHGKKVLGAFSDGLAHLDNLKGTFATLSELHCDK
LHVDPENFRLLGNVLVCVLAHHFGKEFTPPVQAAYQKVVAGVANALAH
KYH

In regions of Africa where malaria is common, the sickle-cell mutation in beta globin is common. A single base-pair alteration in DNA of the beta globin gene leads to the change of a single amino acid in the beta globin protein. A replacement of an A by a T causes replacement of glutamic acid by valine (an E to a V in the single-letter code, the mutated E is shown in bold). This results in reduced haemoglobin function and increased resistance to malaria for individuals carrying one normal and one altered form of the beta globin gene. Individuals with two sickle cell mutations suffer from sickle cell disease. The red cells, which carry haemoglobin, become distorted and this can lead to blockages as the red cells circulate the body. Under conditions of stress, individuals with sickle cell disease can suffer from crises that include extreme bone pain.

The major achievement of human genetics in the past decade has been the discovery of the underlying genetic alteration in many of the simple genetic diseases, devastating diseases such as Huntington's disease, cystic fibrosis, Duchenne's muscular dystrophy, retinitis pigmentosa, myotonic dystrophy and many others. These diseases have a terrible impact on afflicted individuals and affected families; however, they are a relatively uncommon cause of disease. Most of the disease burden on our society is due to diseases caused by the interaction of many genes with ill-defined environmental factors. These diseases include many cancers, diabetes, heart disease, arthritis, depression, schizophrenia and the afflictions of ageing. The challenge for the next decade is to unravel the causes of these diseases. The reagents we need are the complete sequence of the human genome, a catalogue of all the variation in sequences found between different individuals (the common sequence variants will number about three million), a method for detecting the three million sequence variants, and a careful description of the variation in disease symptoms in our population.

The tool that needs to be applied is genetic epidemiology.

Let us consider the disease arthritis. I propose that all the people in Britain with arthritis should be placed on the Isle of Wight and all the people without arthritis should be sent to the mainland. Those on the Isle of Wight must now share the genes and environmental experiences that result in arthritis. A careful comparison of differences between the two populations should reveal the underlying genetic and environmental causes of the disease. This experiment would be inconvenient for both those who currently live on the Isle of Wight and those with arthritis. A virtual segregation and comparison would be just as effective and could be applied to any disease. This is essentially the method that Sir Richard Doll used to link cigarette smoking with lung cancer in the 1960s but with an added consideration of genome sequence variation.

A recently formed consortium between the Wellcome Trust and ten pharmaceutical companies (the SNP consortium) is identifying the sequence differences between individuals and placing the information in the public domain. There are several high throughput methods, such as microarrays and mass spectroscopy, being developed for testing for sequence variants. The only thing missing is the description of variation in disease symptoms for each individual. This should be the job of the National Health Service. Considerable expenditure has been incurred to computerise general practitioner (GP) and hospital records. At the moment such records are not combined despite the very obvious patient benefits to be gained: if you were admitted to any hospital, your medical history could be instantly available to any attending physician, unnecessary duplication of medical tests could be avoided and diagnoses based on complete records would be more accurate. The same information could be the starting material for epidemiological studies. The story would be complete if we were to bleed every person in the UK and produce DNA for typing sequence variation.

The advantages to the UK would be very large. Patients would benefit from improvements in diagnosis and treatment (see below). The NHS would become a major contributor to medical research and would drive outcomes-based medicine. The alternative is to allow the new technology be developed by others and the UK can buy new medicine from the rest of the world.

It has been claimed that the collection of medical and genetic data may lead to abuses in employment, insurance and education. Unfortunately, humans need little excuse for discrimination; sex, sexual orientation, colour of skin, the place of birth of your grandparents and the football team you support have all been the cause of social exclusion. Except for disliking people who support Manchester United, this is clearly stupid. Nevertheless, new excuses for discrimination have to be taken seriously.

### 'The Doctor's Dilemma'

The sequence of the human genome will help us to understand who we are and why we get disease but the big pay-off will be to improve human health.

Imagine you feel unwell. You try taking a couple of aspirins but you still feel unwell so you decide to visit your GP. After waiting the requisite hour after the time of your appointment, the doctor has a few minutes to share with you. The list of your symptoms is explained. The doctor takes your pulse, measures your blood pressure and peers into five of the seven bodily orifices. At this point the doctor is faced with a dilemma. He or she must decide on a treatment for your condition. Three options are available. The first option is to provide social support but not to offer medical intervention. The second option is to refer you to a colleague who has a very sharp knife and uses it to re-arrange your anatomy – this is the surgical option. The last option is to offer you a drug – the pharmaceutical option. I am in the drugs business, I try to make drugs that improve people's health but I have to admit that matching the right drug to the right patient is not an exact science. Many commonly prescribed drugs do not help all of those prescribed the drug. One reason for this partial failure is that the definition of disease is frequently inaccurate. When the underlying cause for a disease is not known, the disease is defined by the symptoms. With a better classification of disease, it should be possible to only prescribe a pharmaceutical to those who will benefit. Even more worrying is the number of people who suffer an adverse reaction to taking a drug.

In 1994, adverse drug reaction was the fifth leading cause of death in the USA; more people died from taking prescribed drugs and OTC medicines ('over the counter' pharmaceuticals such as aspirin) than died from either accidents or AIDS. Although before permission is given to market a drug, it has to be demonstrated that the overall risk–benefit ratio is positive, this is not much comfort to individuals who suffer adverse drug reactions. There are two reasons for these unpleasant outcomes. The first is that the poor classification of disease results in some patients being given drugs that are not targeting the disease the patients are actually suffering from. The second reason is that genetic sequence differences between patients mean that different patients may respond differently to the same drug. As described above, the solution is obvious. The fruits of the Human Genome Project combined with an epidemiological approach to the NHS could lead to a better classification of disease. Exactly the same approach can also be used to define the genetic components of drug response and of adverse reactions. We would all benefit if we could make the NHS a test bed for the new approach to medicine.

### The future

I predict that the genome of every child will be sequenced during pregnancy. The sequence will be used to look for new mutations not shared with the child's parents and to define the precise constellation of variants inherited from both parents. This information will be predictive of threats to future health and well-being. Individual choice will allow preventive modification of the environment and early medical intervention before disease occurs.

The path to the future will provide many complex ethical problems and the always present threats of totalitarian solutions. Despite all evidence to the contrary, I believe that humans are capable of rational thought and behaviour. We can make the world a better place for us all to live and flourish.

FURTHER READING

Goodfellow, P. N. and Lovell-Badge, R., 'SRY and sex determination in mammals', *Annual Review of Genetics*, **27** (1993), 71–92.

Special issue, 'The human genome issue', *Nature*, **409** (2001), 745–964.

Special issue, 'The human genome issue', *Science*, **291** (2001), 1145–1434.

Watson, J. D., *The Double Helix: A Personal Account of the Discovery of the Structure of DNA*, with a new introduction by Steve Jones, London: Weidenfeld and Nicolson, 1997.

# 3 The Bioethics of Reproduction. Have the Problems Changed?

MARY WARNOCK

### Introduction

It is perhaps timely to have a new look at the moral issues surrounding human reproduction, it being more than fifteen years since the publication of the Government Report on Human Fertilisation and Embryology, twelve years since the passing of the Human Fertilisation and Embryology Act in the UK, and nearly twenty-five years since the birth of the first test tube baby, a landmark that led both to the setting up of the committee of enquiry and to the Embryology Act. It is thus reasonable to look back over those years, and see to what extent the ethical issues have changed. There might in theory be two kinds of change. First, the attitude of society to the issues might have altered; secondly, new techniques might have been developed, giving rise to new ethical problems. Up to a point one could say that both these kinds of change have come about.

### The involvement of philosophy

Before addressing the question of these changes directly, however, I would like to say a word about the involvement of philosophy in the issue. I need not remind you that during the second half of the twentieth century there was a huge explosion of interest in, and research into, microbiology and genetics. Every university can testify to this, and every school has seen a vast increase in the numbers of those studying biology, with the express aim of entering this field at undergraduate level. New frontiers of knowledge are being opened up, and, as always happens when there is a revolution in science, philosophy is sucked in. The most fruitful work in philosophy has nearly always arisen in response to challenges from science. Descartes, for example, the first truly modern philosopher, had to try to bring together a God-centred way of thought with his new-found belief in universal mechanical causation, and the newly empirical methods of science. Kant, perhaps above all philosophers, developed his whole vast critical system to try to fit human knowledge and human morality into what he regarded as the absolute certainties of Newtonian physics. It is not surprising, then, that the new biology should have given philosophers new things to think about, both with regard to the very concept of a human being, and, closely related, with regard to the morality of new kinds of therapy and new forms of reproduction. Just as in the time of Descartes, people deeply feared that the security of faith would be eroded by the new science, so it is plausible to suggest that lying behind the moral questions raised by the new biological knowledge, and especially the new technologies that have been developed, there is widespread fear. The same kind of fear of where science might lead was experienced by many of us in the 1940s and 1950s when we first learned of the existence of nuclear physics. The question many people ask now, in connection with biology, is the same as we asked then in connection with physics, namely is it right to do all that can be done? Are scientists the right people to decide where to go next? The difference between now and the 1950s is that now people are prone to look to philosophy to help them to answer such questions.

### Biological mythology

It is perhaps worth noticing that, even more than is the case with the physical sciences, the biological sciences are, for ordinary non-scientists, the subject of powerful myths, from Mary Shelley's *Frankenstein* to Aldous Huxley's *Brave New World*, both myths being frequently invoked in the expression of the public fear of where science might be leading, to the extent, indeed, that merely to mention these stories often seems enough, without argument, to condemn science as leading to effects too terrible to contemplate. For the religious, moreover, the concept of God being particularly bound up with the beginning and the end of life, with man made in His image, and life as a divine gift, the new biotechnology seems in some way presumptuous, taking on the proper role of the deity. It would be impossible to count the number of times biological scientists and the medical profession have been accused of 'playing God'. Once again cliché has become a substitute for an examination of the facts. It is perhaps part of the role of philosophy to attempt to persuade people to think about the facts and their implications without having recourse to mythology or metaphor. This was as true twenty-five odd years ago as it is today. In this sense, the problems have not changed. Philosophers do not, or should not claim to know what is right; they may, however, try to help in distinguishing fact from fantasy, and in preparing, where it is necessary, for a new look at the world.

## The status of the early embryo

To return to the Embryology Act of 1990, we may find that there have been certain changes in public perception since the passage of the Bill through Parliament, changes that were in fact beginning to be evident during the passage of the Bill. The Bill was in part designed to reassure the public that science could be regulated, and that Parliament, through legislation, was capable of blocking where it chose the slippery slope down which so many people feared that we should slide, if once *in vitro* fertilisation (IVF) were to become a recognised, even routine, form of treatment for certain kinds of infertility. IVF involves fertilising a couple's sperm and egg in a dish (or 'test tube') in the laboratory, and then placing the fertilised embryo, or more than one, in the uterus of the female partner. The embryo live in the laboratory outside the human body was a wholly new object at that time. And so the most controversial and difficult issue that had to be debated, first by the Committee of Inquiry and then by Parliament itself, was the status that ought to be accorded to this live and newly fertilised human embryo. Was it to be permitted to use some of such embryos for the purposes of research? It was generally, though not universally, recognised that, in order to improve the success rate of IVF, at first very low, it was necessary to carry on with the research that had preceded the first successful IVF birth, using embryos fertilised in the laboratory and then destroyed, This research was necessary both to learn more about the process of fertilisation and the development of the very early embryo (about which surprisingly little was known at this time), and to find, by trying, the best medium in which fertilisation could successfully be brought about in the artificial environment of the laboratory. If such research had been prohibited, no further IVF programme could reasonably have been carried out, since it would manifestly be morally wrong to offer a form of treatment with such a low expectation of success; indeed to continue to offer it would in effect be to use hopeful mothers as research material. In attempting to answer the question whether or not any research using human embryos was to be legitimate, Parliament was thus deciding whether IVF was or was not to be a permissible treatment for infertility. It was morally wrong to say, as the Irish were inclined to, 'We'll do the treatments, but someone else, (you), can do the research'. It became clear in Parliament, as it had been clear among members of the Committee, that those people who believed that all human life, at whatever stage of development was equally worthy of protection (or equally sacred, as it was often put) could not agree to the use of embryos for research purposes, even at the embryo's four-cell stage. Arguments were not effective with those who held such views, for they were held as a matter of basic principle or dogma, not to be questioned, and immune to argument, especially the consequentialist argument about the good to the infertile, which must be weighed against the harm to the embryos. No demonstration of the advantages that would flow from the continuation of research could move such people. Nor should they have been moved if what they held to be true was in fact true, namely that there is a moment at which human life begins, or when the human soul enters the body, and that this moment is the first division of cells; and that from this moment the living human must be treated with the same degree of respect and protection as is accorded to children and adults. What was difficult to explain to such believers was that their belief, however strongly they held it, was not a scientific belief, nor indeed a belief founded on any facts. It was a judgement of value. This value judgement was based on another, namely the very generally shared belief that human beings have a peculiar value, and must not be wantonly destroyed. The difference between the believers and the scientists was over how to regard the very early human embryo: was it sufficiently like a human person to be accorded the same moral status as a child or an adult? (Of course in a way this was a familiar enough issue, arising with regard to the more developed fetus, in the case of abortion.)

## When does life begin?

Over and over again in the course of the discussion on the Committee of Inquiry, and in the later debates in Parliament, people raised and re-raised the question 'When does life begin?'. Some even said that we should have a moratorium on all research using human embryos until such time as scientists could better answer this question. But of course the question was wrongly posed. It seemed to be a question that might have a factual answer, but it was in reality a different kind of question. It was concerned with when a human being is to count as a morally significant entity; this is not a factual or scientific question, though science and an understanding of the facts of embryology might contribute to a sensible answer to it. Moreover the question was misleading in another way. There is no dramatic moment when life begins. The egg and the sperm are both of them alive, and indeed both of them human, if this means that, in the cases with which we were concerned, they do not belong to any species other than the human. It was the most difficult thing in the world, both in Parliament and in the Committee that preceded the legislative debate to get people to accept that what we were concerned with was a value judgement: how should we value the very early embryo? Should we or should we not accord to it the same status as that which we accord to a developed human being? A particular status is not a natural property like a particular size; it is something perceived as appropriate or granted to some object, doubtless on the grounds that the object has certain properties, and not just arbitrarily, but none the less distinct from those properties. Once the question of the status of the early embryo was recognised as a question of value, not of scientifically establishable fact, it could be accepted that differences of view were inevitable, and that, as far as public policy as opposed to private moral feeling was concerned, we would have to try to reach a compromise, a solution to the problem that could roughly represent a consensus, which was, in short, 'acceptable', even if not exactly welcome, to the majority. Such a solution was reached in Parliament with enormous help from the then Lord Chancellor, Lord Mackay of Clashfern, who dispassionately summed up the arguments on both sides, for and against the permitting of research using human embryos, and, by including two alternative clauses in the Bill, one for and one against, enabled people to vote on this clear issue. The other important source of help was the then Archbishop of York, John Habgood, who had begun his career as a biologist, and who explained a doctrine of Darwinian gradualism, rather than any sudden beginning of the human person. Christianity, he said, no more requires us to believe that human life begins at a certain moment than it requires belief in the Garden of Eden. Fundamentalists were shocked, and inclined to the view that the Archbishop was not a Christian, but Parliament as a whole was persuaded. It was accepted that the use of human embryos for research should be regulated, not prohibited. A gradualist view of the sanctity of human life was adopted.

### A cut-off point for research

An absolute cut-off point for research was to be fourteen days from the fertilisation of the embryo, this well preceding the time when it had even a rudimentary central nervous system, and thus well before the time when it could possibly experience pain, or indeed anything else. It also coincided with the last time at which an embryo could split spontaneously, so as to result in two embryos, identical twins and with the time at which the cells of which the embryo was composed ceased to be totipotent, i.e. became specifically functional. Thus before fourteen days from fertilisation one could plausibly argue that there was, so far, no single individual whose cells had been separated from those of the placenta, or who was indisputably one potential individual fetus. If you think of yourself now, as an individual, you cannot really trace yourself back to the pre-fourteen day embryo. For you are not identical with that; it might have become two. The argument was not that human life began at fourteen days, nor even that at that stage a human embryo became a human person. It was rather that it was deemed legitimate to value the developing embryo differently before and after the development of the primitive streak (the tissue from which the central nervous system of the individual would be formed) and the rapid changes which came after that time, approximately fourteen or fifteen days after successful fertilisation, itself a process rather than single event. To keep an embryo alive in the laboratory for more than fourteen days (not counting any days when it might be frozen) became a criminal offence, carrying up to ten years' imprisonment as penalty. There would be no Frankenstein's monster to emerge from the laboratory; and so, it seemed, public fears were more or less laid to rest. But it has to be remembered that there were, and are still, those who believed that human life, even at its outset, should be protected. However, the regulation of research afforded by the new law became more and more apparently acceptable. The decision of the Committee of Inquiry, later incorporated into the law, to impose this fourteen day limit has been described as drawing a line in sand. If this implies that the fourteen day limit cannot be expected to last, then I am not sure that I think the metaphor apt. But if it implies only that it was essential to draw a line somewhere, however shifting and ill-defined the medium, then I accept it. Such perhaps less than satisfactory line-drawing is often necessary in order to bring various moral and religious beliefs to an acceptable legal consensus.

## The clarifying function of philosophy

All this is history. I have recounted it at length only to show the way in which philosophy may attempt not so much to answer questions as to clarify what sort of questions they are, and whether a new way of posing them may not lead to a possible solution. Once limited and regulated research was permitted by law, those working in this field knew where they were, and could proceed to make IVF a more successful procedure. It is now widely accepted, in this country and elsewhere as an almost routine procedure, in some types of infertility, though its success rate is still not spectacularly high.

## A change of emphasis

There was a marked change of emphasis and of interest, even during the time when these issues were being debated in Parliament (and one must remember that six years passed between the publication of the Report of the Committee of Inquiry and the passage of legislation). Increasingly, in arguing in favour of permitting the use of embryos for research purposes, people concentrated not on the benefits to the infertile, but on the benefits to medical knowledge as a whole. The passage of the Embryology Act coincided with the explosion of interest in genetics, and the establishment of the Human Genome Project, whose aim was to map all human genes. For the first time it seemed realistic to suggest that, using the techniques of IVF, and either by pre-implantation selection of embryos or by pre-implantation gene replacement, some of the most horrendous monogenetic diseases could be avoided. In the House of Lords, the newly appointed Lord Walton of Detchet made a speech that had great influence in securing the acceptance of the Bill, in which he spoke movingly of the benefits to be gained from research using embryos, for families likely to have children suffering from Duchenne's muscular dystrophy. It was such possibilities as this that persuaded many people that embryo research was medically imperative; and of course, as I have said, such medical developments took the future of IVF for granted.

### Future problems within fertility treatment

Before discussing this change of emphasis further, I want to say a few more words about IVF itself and related treatments, to suggest that even here there may be unforeseen moral problems. The questions that are most often raised now are not whether the treatment itself is morally acceptable, but rather who merits treatment. The issues here are concerned with social rather than strictly medical morality. Yet they arise directly out of the new medical technologies.

These were problems, which the Committee of Inquiry, as I now see it, failed to deal with satisfactorily, which, in fact, they appear to have fudged. Perhaps understandably, we were less than sure-footed in discussing future social policy. We addressed but never properly answered the question whether some couples, or some single people, seeking treatment by IVF would be turned away. We invented numerous hypothetical cases to consider: potential parents with a record of child abuse; homosexual couples, male or female, who wished to bring up a child; women who had given birth to children but who had been sterilised, and who, having married again wanted a new family. Because our original remit had been narrowly defined as the treatment of infertility, we were ill equipped to consider the new techniques in relation to people who were not infertile, but wanted to make use of these techniques for various different reasons. We fell back frequently on the solution that the doctors involved would find a clinical basis on which to decide the issue, or, worse, that the would-be users would be subjected to counselling. I remember an occasion when the Committee was visiting Northern Ireland, and I asked a gynaecologist from Belfast how he would proceed if a couple strongly suspected of child abuse came to him seeking IVF treatment. His reply was that he would counsel them. When I asked what he would do if, none the less they persisted in their request, he said he would counsel them and counsel them until they went away.

In those far-off days we assumed that IVF would be a treatment readily available on the National Health Service, the NHS. Now many NHS clinics have been closed and, increasingly, IVF is a treatment available only to those who can pay. It was relatively easy to lay down that an NHS facility should be available only to the genuinely infertile (though there would still remain ethical problems). But now there are private clinics, especially in London, willing to carry out IVF and other procedures for anyone who can pay. Such clinics are subject to regulation and inspection, and are governed, like NHS clinics, by the 1990 legislation. Often their record of successful treatment and properly conducted research may be good, and they therefore have no difficulty in obtaining a license from the regulatory authority. The Human Fertilisation and Embryology Authority (HFEA), set up as a central body by the 1990 Embryology Act, has an ethical as well as a scientific and medical responsibility, and it may, though with some notable exceptions, tend to leave the question of who is to be treated to the Ethical Committee of the hospital or clinic itself. In the private sector, such ethics committees may not be wholly free from suspicion that they have the financial health of the clinic rather too close to the front of their minds. When the Committee of Inquiry first visited the famous clinic at Bourn Hall in Cambridgeshire, we found that an ethics committee had been set up just before our arrival, and consisted of the doctors involved and the accountant, with a lay member recruited from the village. Even if things have vastly improved since then, it is necessary to reflect that, the more IVF and related treatments fall into private hands, the more commercially motivated they are likely to become.

To illustrate the kinds of problem that may arise, let me remind you of an outcry in the press over a couple who decided to have an embryo fertilised in the laboratory of a private clinic and frozen for a number of years, so that the mother, a successful business woman, could have it implanted when she was 40, i.e. when she thought she would be ready to have a family, thus avoiding both the immediate disruption of her career, at a critical point, and some of the risks associated with late conception. Another much-publicised case involved the determination of a woman to have a baby after her husband had unexpectedly died young and she had caused sperm to be extracted from him when he was in a terminal coma. This last case was ruled illegal by the HFEA. But the woman was permitted to take the sperm to Belgium, where artificial insemination could be carried out legally, and the baby has now been born.

It is certain that such services, if energetically enough sought, will be provided in other countries, even if prohibited in the UK. All that is needed is money. For my own part, I doubt whether it is worth trying to prohibit by law that which increasingly the market may demand. For the market will inevitably be small, and the costs will be high. Most people will continue to have their babies in the regular, and affordable, way. We may wish that the few others would not go to such extreme lengths to have children, or to have children just when they want to; but, as far as social policy is concerned, it is likely to remain a limited problem. Of course one might argue that there are issues of equity. Why should the rich have access to better, or rather more complex, services than the poor? But this, alas, is a general problem for the NHS. Even straight infertility treatment has come to seem a luxury that it is difficult to supply to everyone in need, free from charge.

### New techniques, new problems

So what are the new problems, arising not from our increasing familiarity with IVF and our increasing assumption of its success but from new technologies? These, I have suggested, focus on the increased knowledge of the human genome, and the possibility of ensuring that, without having recourse to the abortion of affected fetuses, babies are not born with inherited monogenetic diseases. This may be either through the technique already available of pre-implantation selection from among embryos fertilised in the laboratory, for parents at risk of having affected babies only healthy embryos being selected for implantation, or, in the future of pre-implantation or prenatal genetic manipulation, to replace or remove an identified faulty gene. It has to be said, however, the latter looks more and more like science fiction as time goes on. The development of such manipulation, confidently asserted in 1990 to be about five years away has in fact been very slow to take place.

I personally find it difficult to see any moral objection to embryo selection, i.e. a programme of fertilising *in vitro* a number of embryos from couples at risk and selecting for implantation only those embryos which are free from the disease in question, or implanting only female embryos where only male children are sufferers from the disease. Such treatment, in my view, ought to be available on the NHS to couples at risk; it could be argued that money might be saved thereby in the long run.

## The disability lobby

There are those, however, who increasingly raise objections to such programmes on the grounds that to choose that only healthy babies should be born is to diminish or cast doubt on the value of other people who have been born with genetic diseases, or even of people with acquired disabilities. The most extreme of such objectors argue that disabilities are not actual, but are constructed or perceived by society: it is society, not physiological or genetic impairment that disables people; a couple who prefer to give birth to a 'normal' child ('disability' and 'normality' always being placed in inverted commas) are guilty of a prejudice known as 'ableism'. I cannot think of many parents who would not prefer to have a healthy rather than a sick child, for example a child who did not suffer from a life-threatening disease such as Tay–Sachs disease or Duchenne's muscular dystrophy, and this not for their own sake (though doubtless this as well) but for the sake of their potential child, whom they already love and cherish, before his birth. That there are some deaf parents who would prefer a deaf child, or some dwarf parents who would prefer that their child was a dwarf, does not, I think, constitute an argument against this general proposition. Nor am I sure that parents who belong to their particular 'disabled' culture, and who want their children to belong to it alongside them, are being entirely disinterested with regard to their children. For there is all the difference in the world between being profoundly deaf, say, and becoming part of the deaf world, and, in contrast, suffering from a devastating disease that entails a painful, miserable and short life. I cannot accept that there is at work here a kind of prejudice like that which makes some parents irrationally hope for a male rather than a female child (though this itself is not irrational, if the inheritance of a property or a title is involved), or a kind of latent discrimination against those who have been born with or have acquired a disability. If medicine, with the help of the new genetic understanding we are acquiring, can either prevent children from being born with genetic diseases or create genetically engineered drugs to treat those not at present capable of treatment, then I cannot see this as other than a benign advance in medicine, just as the discovery of insulin as a treatment for diabetes was. We should not allow ourselves to be browbeaten by the lobby of the existing disabled into either denying that they have disabilities or giving up the search for a remedy for future generations of children.

## Cloning

It is impossible to keep up with or list all the new discoveries in the field of genetics at the present time. But none has had more impact, nor thrown up more apparent moral problems that the production, by cloning, of Dolly the sheep, and Dolly's successors, Molly, Olly etc. On 7 March 1996 the periodical *Nature* carried an account of the attempted cloning of a sheep by a new technique, at the Roslin Institute, near Edinburgh. A year later the same periodical had the story of the one successful live birth resulting from the experiment (numerous attempts having been recorded), of a lamb known as Dolly, now fully mature and having had lambs of her own. The birth of Dolly, without a father, from the extraction of a cell from an adult female sheep caused a great sensation in the media. Meanwhile other experiments had been taking place, notably the cloning, by a different technique, of monkeys at a private research centre in Oregon. Mice have also been successfully cloned. The inevitable question was raised: if other mammals can be cloned, why not humans?

Apart from the present low success rate, which might be expected to improve over time, what was to stand in the way of the cloning of humans? The offspring, like sheep, could be born with chosen genetic qualities from the tissue of another adult human. So insistent was this question that the HFEA issued a Consultation Paper to find the reaction not only of scientists but of the public at large to the prospect of human clones.

### The fear of human cloning

It is perfectly obvious that the cloning of human beings is, or will be, possible. Why do people react so immediately and so strongly against the thought? The first thing to be said, which has often been said, is that we are perfectly familiar with and accustomed to genetic clones. We do not react with horror or distaste from naturally formed identical twins, whose genetic make-up is literally identical, one with another. Indeed identical twins, as we are now often reminded, are in fact genetically closer to one another than Dolly was to her originating 'parent', who is of course not so much a parent as a sibling born a generation apart. For identical twins share all of their genes, whereas Dolly was the result of the nucleus of a cell from her 'parent' being placed in the outer membrane or cytoplasm, of a cell whose own nucleus had been removed. The cytoplasm contains a few, but significant, genes of its own, specifically those in the mitochondria, mutations in which can cause horrendous diseases in humans. Dolly, therefore, contained some genes that came not from the 'parent' from whom the tissue was removed, but from the donor of the egg that received the nucleus from the 'parent'. Strictly speaking, then, she is not identical with that 'parent'.

In any case, since we recognise that identical twins, or natural clones, are distinct human individuals, in no way deprived of freedom, worth or personal identity, it cannot be genetic identity itself that appals us. There must be some other source of the fear that grips people when they think of human cloning.

There certainly seemed to be some odd features of Dolly when she first came into existence, which may have been part of the source of fear at first. But most people did not understand enough of the details of her birth to have been much alarmed. She came into the world with no father, as the result of a pregnancy started by the removal of tissue from an adult female. Some people, I suppose, might be alarmed by the fact that this could in theory do away with the need for a father in the life of a child. Another oddity of Dolly was that parts of her were of a different age from other parts. The nucleus which came from the 'parent' was adult, but her mitochondria gave her the status of a neonate. It was not at first certain how she would age. However, apart from a degree of cellular ageing known as telomeric shortening, she does not appear to have suffered from this collection of oddities, and has already given birth to her own lamb. (There is, however, accumulating evidence that other cloned animals suffer from fairly frequent gestational abnormalities.) The main source of anxiety, however, was the degree of intervention that was necessary to make a clone, the numerous failures that preceded the one successful birth, the possibilities of human error in the process, and the number of women who would be involved in the experimental attempts to produce human clones. These factors make it impossible that the cloning of whole human embryos should be undertaken at the present time, despite wild promises that people will be able to be cloned within a few years.

Some people have welcomed the birth of Dolly on the grounds that it makes possible (or will make possible when techniques are improved) the quick and reliable generation of improved farm animals: here, I believe, we are coming near to the source of the fear of human cloning. For if this is so, then equally we could see the quick and easy generation of supposedly 'improved' human beings. And this raises the question of how do we know what an improved kind of human being would be? Who decides what is the best? Are we to think of some neo-Aristotelian set of characteristics of what counts as the best kind of person to be? If so, who decides on the criteria? Or, more alarmingly, are we thinking of politicians who might wish to create people who would be trouble-free, useful, docile, satisfactory at generating wealth and not generating criticism?

### The fear of genetic determinism

Here lies the root of our fear; it is part of the general fear we have of being used or manipulated for someone else's ends. The belief that each human being is of infinite value, intrinsically and for himself, is bound up with two further beliefs. The first is that how we are, individually, is partly a matter of chance, of the mixture of genes we happen to have inherited from our parents, and no one can wholly predict this, still less control it. The second is that, whatever genes we have inherited, we are able, up to a point to choose how we live, whether to improve ourselves, whether to rebel against our parents and our background, or whether to accept it and incorporate it into our chosen way of life. We need to ask how much of this sense of freedom and individuality would be lost to us if we were born as the result of cloning.

Certainly the element of chance would be much diminished if we were artificial clones. Apart from the small number of mitochondrial genes we would get from our egg-donating 'mother', we would have none but the genes of our 'parent' (that is our sibling of a different generation). We would not have the chance mix we have as a result of regular fertilisation and birth.

On the other hand, whatever genes we had, we would be as much or as little free to choose our way of life as we are with our chance mixture. For we are not nothing but our genes. We would still, as we are now, be shaped largely by the chances of our environment, not just our physical or geographical environment but whom we happened to meet in our lives, whom we met at school, who taught us, where we went on holiday, etc. All the things that go to shape us now would shape us still. Are we rational, then, to be so much horrified by the idea of human cloning?

The answer may well be that we are not (and this is quite apart from the fact that the risks and expense are such that it is unimaginable that human cloning should be undertaken in the foreseeable future, even if it is possible in theory). For there is a further point to be made. The fear of human cloning may best be seen as the fear that somehow someone will become so politically powerful as to treat human beings like cattle or racehorses and hope to produce the best breed of humans, as he or she might try for the best breed of cattle or of horses. But in fact to produce even cattle or horses by asexual cloning would be a deeply mistaken policy. The diversity brought about by ordinary and chancy reproduction will always be preferable, even among cattle and racehorses, even though there may seem to be one and only one function that these animals are supposed to fulfil. The diversity of the gene pool is a safeguard against the development of genetic defects that might lead to the weakening, even the destruction, of the whole species. It is doubtful, therefore, whether anyone would be so foolish, even if they could conceivably become sufficiently powerful, to embark on large-scale compulsory cloning of humans.

### Could there be a use for the cloning of whole humans?

If, then, we can put this great fear from our minds, it remains to ask whether there are any circumstances in which reproductive cloning (cloning that is, of whole human beings, rather than of pieces of skin or other tissue) might be a chosen way of birth, not for large numbers of the human race, but for the occasional individual. It seems possible that there might be certain kinds of male infertility where the nucleus of a cell from an adult male might be used for merging with or placing in the cytoplasm of an egg, which would then be placed in the female partner's uterus, whether she or another had provided the egg. The objection to this process, as I have suggested is the extreme risk, in the present state of the technology, to any child that might result, or the risk to the mother, in the case of repeated failures. Since every species of mammal is different, no number of experiments with sheep or mice would seem, at present, to reduce the risk to humans to an acceptable level. The degree of risk, therefore, seems of itself to constitute a moral argument against whole human cloning.

However there is still an enormous amount that is not known about the early development of differentiated cells, and the parts that the nucleus and the cytoplasm jointly play in the development of the embryo. Research into these matters is likely to be of great importance in finding ways to combat genetically inherited diseases. Since 2001 it has been legally permissible in the UK to fertilise human cells by cloning, whether for the purpose of observation only, or for the production of drugs or tissue. Any embryos resulting are subject to the fourteen day rule; but much can be learned in this time.

A number of countries, immediately after the birth of Dolly, brought in legislation either to prohibit human cloning or to state that no experiments concerned with cloning would be funded out of the public purse. Others have tightened existing legislation. The more successful cloning of other mammals is carried out, the more likely it is that legislation will be introduced, for instance by the European Community. However such legislation should be introduced with a realistic eye to what the dangers actually are. No one I think could reasonably doubt that human cloning, if it ever becomes possible, should be subject to regulation, in order to rule out and criminalise the mad dictator, or the mad scientist, or even the quack doctor, who might promise to the gullible, for a fee, more than he or she could possibly deliver. Yet the possible increase in knowledge and the advantages to science and medicine should not be overlooked, nor forgotten in a rush of panic.

### How far ahead should we look?

The new possibilities of genetic manipulation, not cloning alone, undoubtedly give rise to new moral problems, barely thought of in 1984 when the report of the Committee of Inquiry was published, though we did have a short chapter on possible developments beyond the issue of infertility treatment.

The major problem, then and now, is how far ahead it is reasonable to try to look. Should we issue guidelines or embark on legislation to try to control, in anticipation, developments that have not yet taken place, but which, in theory might take place within the next decade? This is in itself a moral issue. Shall we be blamed by our successors and descendants either for letting things slide, allowing scientists to descend further and further down the dread slippery slope, keeping our heads in the sand, or shall we blamed, especially by scientists and the medical profession (but also by possible beneficiaries of new techniques) for binding our successors too tightly in response to our own timidity? Should we now be drawing up a list of monogenetic diseases, in the case of which alone genetic screening and gene therapy should be permitted? Or should we allow that, when such things are possible, every-one may choose, if they can pay for it, that their babies do not have the genes they wish them not to have (or even have the genes they prefer). Should we be so frightened of 'eugenics' as to forbid all research that might lead to the replacement of genes? Again, should we continue to hold, as most countries do, that there must never be manipulation, for whatever purpose, of germ cell genes, but only of the genes of somatic cells? These are all moral questions for the future. But the immediate question is whether we should prepare ourselves for this future by restrictive legislation. It may of course be made a question whether, in the increasingly international world of science, medicine and pharmaceutical research and development, it is worth while or effective to legislate, one country at a time. Are we doing any good by declaring that there are some things that will not be permitted in our backyard, when across the Channel or at least the Atlantic they may be perfectly legal? Perhaps our efforts should go into trying, globally, to pre-vent the exploitation of the new technologies without proper trials, and for inordinate gain.

Just at the moment my instinct is that we are not ready for new legisla-tion; but we need more discussion and more access to information on the rapidly changing scene. But in the nearish future perhaps there should be a standing Royal Commission set up, like that on environmental pollution, which would have as its task to keep abreast of issues as they arose, glob-ally, and report to Parliament. Then these issues could be debated with a view to deciding whether or not new legislation was required. The Nuffield Committee on Bioethics has nearly performed this function for years. Perhaps it is time for an equally independent body to replace it with sufficient resources to take on both a watchdog and an educational role. It is of the greatest importance that information should be readily accessible on any new developments, but information is not enough. We need understanding. I believe that at both school and university students should be taught to take a dispassionate and long-term view of the problems in bioethics; that there should grow up a body of good scientific journalists who could avoid the dangers of scare-mongering or sensationalising matters that are crucial to many individuals, and also to society as a whole. For the only thing of which we can be quite certain is that the problems in this field will continue, as knowledge increases, and that they will not go away.

FURTHER READING

Ford, N. M., *When did I begin*? Cambridge: Cambridge University Press, 1988.

Galton, D., *In Our Own Language: Eugenics and the Genetic Modification of a People*, London: Little, Brown, 2001.

Harris, J., *Clones, Genes and Immortality*. Oxford: Oxford University Press, 1998.

Mulkay, M., *The Embryo Research Debate*, Cambridge: Cambridge University Press, 1997.

Rifkin, J., *The Biotech Century*, Victor Gollanz, 1998.

Singer, P. and Wells, D., *The Reproductive Revolution*, Oxford: Oxford University Press, 1984.

Turny, J., *Frankenstein's Footsteps*, New Haven, CT: Yale University Press, 1998.

# 4  The Violated Body

DAVID CANTER

## The colour of murder

Many murderers have similar characteristics. Not only are they likely to be men, but many studies also show they are likely to be in their early 20s, from disturbed, dysfunctional family backgrounds with some prior criminal experience, not necessarily for crimes of violence. These features that murderers tend to have in common can be seen as contributing to a limited portfolio of ways of dealing with other people. Their dysfunctional backgrounds mean that they have difficulty in feeling and knowing what it means to be a person, especially in seeing the world from another's point of view. They see the cause of their frustration, anger or jealousy, or opportunity to slake their greed, as encapsulated in the object of another being. They want to remove or destroy that entity as the only way they can relate to the individual they see as causing their reactions. Murderers therefore provide a rather exaggerated illustration of the consequences of confusing the person and their body, violating the body through acts of aggression, as a product of this confusion.

Yet there are many different ways in which a murder can be carried out. These different styles of murder are likely to relate to differences between the murderers themselves. The vicious rage that leads to violent mutilation is likely to be a product of anger with the victim in which the act of murder is what drives the killer on, the 'righteous slaughter' that the American sociologist Jack Katz identified. In contrast the murderer for profit that William Bolitho identified in the 1920s has the end result clearly in mind as he goes about his devious plan to poison, or kill in some other way, that distances himself from his victim.

Variations in murder and other forms of bodily violation illuminate the confusions surrounding the relationships between the person and the body. These confusions are now at the heart of many debates about the influence that current science may give us over our bodies. Attempts to reduce so much of the complexity of being people to the processes of physics and biology also challenge the rich notions we have of the relationships between self and body. Variations in the chosen modes of killing people might be thought of as the colours that distinguish one dark crime from another. These colours, I will argue, are reflections of many other differences that can be found in the struggle to make sense of being human.

The reason why these various shades and hues of murder are so illuminating is that they do not seem to be created solely by the offender's practical considerations. The use of a knife or a rope, slow poison or an illegal firearm, all derive as much from the offender's lifestyle and ways of seeing the world and his victim as they do from ready availability or the overt demands of the task at hand. The activities surrounding the choice of method for dispatching the victim add further information beyond the murderous act itself. How the body is dealt with after death, for instance whether it is hidden or covered, or the type of interaction that may have occurred before death, such as whether it was a sudden unexpected attack or one that grew out of an argument, also provide important signifiers of the mood and tone of the killing. They may indicate what the victim meant to the offender, such as the role that the victim played in the killer's life. As Bolitho wrote in 1926 about murderers, 'They very commonly construct for themselves as life-romance, a personal myth in which they are the maltreated hero, which secret is the key of their battle against despair'. In other words the victim takes on significance in the offender's self-constructed life story that is reflected in how the body of the victim is violated.

As so often happens, fiction writers appreciated the significance of such matters long before detectives, and in their wake psychologists, began to examine these subjects systematically. In Raymond Chandler's *The Long Good-Bye* the plot revolves around the private detective, Philip Marlowe, being unconvinced that the violent murder with which the story opens could have been perpetrated by the suspect. Marlowe is convinced that the suspect adored the victim too much to assault her in that way. The actions revealed at the murder scene were at variance with what was known about the role the victim had in the life of the person who was thought to have committed the crime. Real life examples are often more horrific but reveal similar processes, as when, for example, the mutilation of sex organs is clearly part of an act of jealousy.

Therefore, beyond the twists it can give to the plot of a thriller, or the assistance to a police investigation, the significance of the variations in violent physical assault raises questions about crucial psychological processes. They draw our attention to the fact that different ways of assaulting the body carry implications for different ways of relating to the person whose body it is. The psychological examination of violations of the body are therefore an important, if somewhat unusual, gateway to considering the fundamental nature of the relationship between the person and the body.

## The range of violations

The problematic nature of violations of the human body, and the profound questions those problems raise, can be illustrated further from something I noticed near the Royal Courts of Justice on The Strand in London. In a telephone kiosk outside the Courts it was difficult to avoid becoming aware that Tara and her colleagues were advertising their services by means of postcards stuck on the walls. What was especially interesting about Tara is that she was willing, presumably for a fee, to be spanked. By contrast, one of her colleagues, who prefers the more anonymous sobriquet Severe Mistress, was charging for the privilege of humiliating, binding and torturing her clients. I suspect that this latter service is rather more expensive than Tara's because of the higher overheads. Severe Mistress boasts a 'fully equipped dungeon'.

The irony of these services being on offer so close to one of the highest courts in the land is that, nowadays, those courts would never countenance spanking as fit punishment for any crime. Neither would they ever endorse humiliation and torture as an appropriate redress for even the most serious of crimes. This irony reflects the changing views we have of our body and what officialdom is allowed to do to it. Torture of many forms was not uncommon in Britain until relatively recently, not only as a method for obtaining a confession and other information but as a form of punishment in its own right. The sentence of 'hard labour' that still obtains in many places is a recognition that the removal of a person's freedom is not enough, but they should have to suffer physically as well. Indeed, the 'boot camps' and 'short sharp shock' that recent British governments have introduced in the sentencing of young offenders is part of a long tradition of severe physical punishments. Many of those used in the past would today be regarded as torture. But it can be argued that there are only differences of degree between some current punishments and the treadmill that was still in use in British prisons less than a century ago, before that the 'cat' that was used to flog British soldiers for minor offences, or the various forms of rack of earlier centuries.

What has happened over the centuries that we no longer hold acceptable punishments that were once commonplace? I would argue that it is the relationship between the person and the body. This relationship has been growing ever more complex over the centuries as the clear distinction between the body and the spirit has eroded.

There are two different trends that these distinctions elucidate. One is the growing view that the person has to be changed by means that go beyond the modification of the body. Another is the growing reluctance to insult the person, even when it is acknowledged that physical punishment is acceptable.

With the growth in the recognition of human identity and personhood has emerged a more psychological approach to torture. Changing the nature of the person through fear and other devices has always been used. But mind-changing strategies have been lowered to new depths in the twenty-first century. Indeed, if a person can be shown to have changed his or her allegiances, and thus, in effect, who he or she is, without any overt indication of physical coercion, that is now deemed more of an achievement than a physical change. In the past the physical mark was seen as a prerequisite of an indication of change, often the ultimate physical control of death.

The distance between Tara and her colleagues and the tortures of the Spanish Inquisition is very great indeed, but it is remarkable what people will inflict on themselves. From time to time suicide occurs in the pursuit of exquisite pain. So it is difficult for us to grasp the huge variations there are between people in their reactions to violations of their bodies. What is clear though, from accounts of the suffering of martyrs as much as from the sadistic and masochistic indulgences of fetishists, is that the role the person himself, or herself, plays in the process is crucial to making sense of their reactions.

People who have been tortured comment on how they lose their sense of identity long before they lose consciousness. A possibly related process of feeling separated from their day-to-day existence seems to be what produces the heightened excitement that seems to characterise some reports of sado-masochism. It is the difference in relationships people have to their bodies that, at its extremes, can make the difference between an act of violation being torture or a service for which people will pay. After all, it is lack of reciprocal acceptance that makes sexual activity rape.

At first sight the difference in circumstance appears to be volition. It is not a violation if you seek it out. But there is a rather more subtle and potentially more important distinction, one that is the foundation for banning many forms of punishment and that is reflected in the American Constitution's Eighth Amendment forbidding 'cruel and unusual punishment'. The concern is not to violate the *person* and the rights that person should have. It is the recognition that the 'person' can be violated even if there is no physical damage. The clients of Tara and others claim the freedom to do to their *bodies* as they wish.

It is the significance of the acts on the body of a person whether self-inflicted or not that raises so many questions about what our body and others' bodies mean and how in many cases the transactions with people are mediated by transactions with their mortal flesh. The caresses and acts of love do, of course, reveal the significance of the persons between whom the actions take place. A touch can be a gentle caress from a lover or a fearful act of gross violation from a stranger. But, the offensive, destructive acts may offer as much of an insight, if not more, into the confusing relationship there often is between our bodies and our selves.

## A scale of violation

The lover's touch, when unwanted, is at one end of a scale of bodily violations. Violent and abusive dismemberment of the body is at the other end. This scale seems to reflect an increasing desecration of the individual as more aspects of their personal, private selves are defiled through the actions on the body. For even the act of touching varies in its significance depending on the body part touched as much as on the person doing the touching. When two strangers have to squeeze past each other in a crowded public place there will be a tendency for them to pass back to back, or side to side. Great contortions will often be gone through to ensure that they do not touch face to face. This shows that it is our faces and the front parts of our bodies that carry so much social and related symbolic significance. These are the parts of our bodies that most capture our unique selves as individuals and are therefore considered most vulnerable to violation.

One interesting consequence of this is that marks to the face have huge significance, as studies of even the smallest facial marks show. A number of psychological experiments, for example, have revealed that quite small facial scars can have a big impact on the judgements people will make of the scarred people. This is also illustrated in the big difference in the practice of tattooing that many young people accept. One current fashion is to have the tattoo on the upper rear shoulder so that it can be revealed or hidden as the person wishes. A tattoo on the face is part of a much more extreme expression of distinctiveness through group membership.

Beyond these forms of apparently minor violations are the wounding and mutilation that people in depressed and despairing states or ecstatic moods inflict on themselves. There are many forms of masochistic acts the world over. Dr Guy Grant, an Australian physician who provides much guidance to police investigations, has been collecting accounts from many cultures of the range and variety of bodily violations people inflict on themselves and others. The extent and nature of these acts are quite remarkable. But what is particularly notable is the range of facial modifications that are carried out for apparently cosmetic reasons. Another large group involves activities relating to sexual organs. Clearly both the face and the sexual organs play a significant role in all cultures in the defining of a person's identity and it is therefore perhaps not surprising that these are popular targets for modification that can lead to violation.

The most obvious extreme form of violation is rape. It is an important question as to why rape is regarded as such a crime distinct from other forms of violent assault. A feminist interpretation could be that the value of a woman as some form of property is greatly reduced once she has been sexually violated. So special laws are required to protect this particular value. It can be seen that such an argument can readily be developed to recognise the special vulnerability of women and therefore the need for them to have special protection. This is particularly worth noting because the crime of male rape has been recognised in many jurisdictions only in very recent times. This is all probably part of the growing acceptance of the particular challenge of sexual violation to many people's identity. Certainly there is growing evidence that sexual assaults of all forms produces psychological confusions and often traumas that go far beyond the intensity of the physical insult itself. Yet again this reveals the very important symbolic qualities that our bodies carry for us.

### Strategies for violation

When we turn to the extreme forms of violation that occur in murder we can see the way the meaning of the victim for the offender is enshrined in the actions committed on the body. As I have hinted earlier, two dominant strategies seem to capture most processes. One is the emphasis on the person, with a limiting of the significance of the body. This can give rise to the mutilation of the body as a by-product of attempting to change the person. It is the person that is the target of the actions.

This idea is open to empirical examination through the rather grisly consideration of what sort of actions co-occur in crimes of violence, particularly those committed by serial killers. This is not an easy empirical area in which to work. The data are hard to come by, partly because of the mercifully few cases available for study. The data that are available will invariably be crude and of low levels of reliability, having been collected by criminal investigators for legal reasons not for the purposes of research. So often details of crucial psychological significance, such as at which stage in the act of murder sexual acts took place, will not be carefully determined or recorded because they have little legal significance. Yet in the Centre for Investigative Psychology at the University of Liverpool we have begun collecting appropriate data sets and some patterns are beginning to emerge that illustrate the processes being discussed here.

The hypothesis would be that when the focus was on controlling and manipulating the person there would be subsets of activities that would share a common theme, or 'colour'. That is what we find. So, for example, sexual activity and attacks to the upper torso, often with little immediately life-threatening implications are likely to co-occur in offences. Some of the victims may even be released. Humiliation and degradation are often objectives because the attack is on the person. The killer here is angry with particular people or what they represent.

A somewhat distinct strategy is when the person is ignored. This gives primacy to the body. The victim is little more than a body to these offenders and its use for their own ends is the driving force that leads them to kill. This is supported by finding that these killers insert objects into the dead body, carry out necrophilia and rituals on the body, perhaps even indulge in cannibalism. This is the psychotic killer for whom the body is independent of any person.

When we have looked at other violent crimes such as rape and sexual abuse of children we find parallels. For example some paedophiles are focused on using children's bodies for their own gratification. These are the violent people who may kill to control or silence their victims. For others it is the childish person that draws their desires. They will spend a great deal of time luring children into an apparently innocent relationship, possibly even believing there is no harm in the acts of abuse they perpetrate.

### The person as invention

In order to understand the further implications of these considerations of bodily violation it is necessary to realise that the 'person' is an invention of the human psyche. Being a 'person' cannot be a taken-for-granted 'given'. That is one of the most challenging implications of modern science. The recognition we each have of 'being me' is not a mere consequence of a corporeal existence, but requires that we each transcend our physical experiences and construct a notion of our selves that goes beyond our bodies.

The central message of many studies of child development, spurred on by the great contributions of Swiss psychologist Jean Piaget and the rich metaphors of Sigmund Freud, is that the crucial stages of early childhood are the distinguishing of the self from others. This starts with the child becoming aware of the separateness of his or her body from that of others that succour it. It then evolves into an awareness of the unique qualities they have as a person. It is distortions in this process that undoubtedly lay the seeds for dysfunction in later life.

What our consideration of the violated body shows is that the notion of self is sometimes a difficult fiction to maintain. It is challenged every time the body is violated in any way. Therefore these violations and the reactions to them can help us to understand more fully the different ways in which we construct ourselves as people and the vulnerabilities inherent in those constructions.

Every one of us takes it for granted that we are a 'person'; an identifiable, unique, sentient human being with a past and anticipated future. Furthermore, except in the most extreme states of mental disturbance, we see coherence in our 'self'. We know, more or less, who we are and what it means to be that person. We do not experience ourselves as animated organisms, as mechanico-physiological systems, or even animals struggling to survive. Even at our most atavistic we regard ourselves as *people* who have certain urges and desires, needs and aspirations.

Yet, evolutionary biology and the invasive insights of biochemistry and neuroscience are making it increasingly clear that this sense of self and the associated awareness of being a person are fictional constructions. Indeed, it is emerging as a by-product of the biological sciences of the past century that the belief we each have in our own identity as people is probably the greatest innovation in evolutionary history. It is a creative leap of human imagination that requires that we minimise both any indications that we are merely sentient animals and all the biological and psychological changes that happen during our lives to turn us into very different entities from those we once were. We have to construct a story about ourselves that encapsulates the central psychological continuity of our existence as the motif around which the variations in our life unfolds. This story is fictional in the sense that it is a particular construction that presents a limited perspective on our selves and in which we are the main characters, carving our identities out of our transactions with the world.

This sense of self and 'personhood' is a far more significant aspect of our experience than the much-studied consciousness. We may, after all, share aspects of consciousness with our close relatives in the animal kingdom. Conscious awareness of our surroundings and even our recognition that we have that awareness and share it with others, including other primates, may turn out to be a natural evolutionary product of a sophisticated cortex. But it requires much more inventive processes, utilising a combination of uniquely human talents, such as language, social interaction and the creation of cultures to produce the firm belief each of us has that we are persons with a special identity as unique beings.

## The emergence of the person

Perhaps the earliest recognition that the body has to be handled carefully because of the person it contains is indicated when early humans buried their dead and made provision for a non-corporeal hereafter. For early peoples the body was an inefficient container for the more important soul. But, as science has dispelled the myths of religion in parallel with giving us more control over our bodies, so the body and the person who enlivens it have become ever more closely equated. Even in these Godless times the care and respect with which the dead are disposed of is a continuing paean to the importance of our non-physical identity. That is why the worst atrocities to be pictured are those of unburied bodies. They challenge our fundamental faith in our own psychological existence.

There may appear, here, to be an equating of the person with other more religious notions like spirit or soul. But the very opposite is my intention. So long as the soul was considered a God-given force that vitalised the body it was feasible to carry out atrocities on the body in order to save the soul. Many of the tortures of previous centuries were supported because of the idea that the immortal spirit of a person was being hampered by the evils inherent in the body. With the demise in the belief in the soul there is a temptation to believe that only the body matters. Its processes and products are seen as the answer to all human strengths and weaknesses. But this ignores the importance of the investment we each make in creating our selves. It ignores the existence of a person that can never be totally reduced to biological and physical processes. It is this recognition of the need to respect a person and his or her identity that has led to the outlawing of extreme forms of torture. It is confusions individuals have in the nature of identity and its relationship to the body that lead to self-mutilation and sado-masochism.

Attempts to modify the living are the obverse of the reflected sanctity of the dead body. Because we cannot shake off the body we must attempt to modify it, in extreme cases violate it, in order to construct it more like the person we want it to be. This takes on an importance far beyond what may be achieved by the practical benefits of nips and tucks, marathon runs or other feats of endurance and prowess. This importance comes from the fact that the body is one of the basic metaphors for all human transactions. Any form of mutilation is thus essentially symbolic. This symbolism grows, in part, out of the very strong traditions that the flesh is profane and it is the spirit that is immortal and sacred.

## Coping with the person – body paradox

The dualism of person and body is therefore not simply a product of rational thinking. It takes on profound emotional significance. For many people there is a struggle between the things they do not like about themselves, as reflected in their body, and what they believe is truly them in the sort of person they are. Utilising the services of Severe Mistress may be one way of trying to cope with this. Other more extreme forms of self-mutilation may provide some temporary relief for more intense inner agitation.

The emotional release of inflicting wounds on the self is difficult for most people to understand. If a person has been abused by people close to them, especially in the early years when they are forming an image of themselves, then there are likely to be a variety of distortions in the way they see them-selves and their bodies. This may be reflected in many different ways, anorexia or bulimia, or, if the individual is in deep turmoil over who they are as a person and what role their body plays in that, in self-injury. The distance self-injury places between body and person can be disturbingly soothing. One person who moved through these experiences into professional life has anonymously posted on the Internet a remarkably insightful and con-vincing account of her personal turmoil:

> At the age of 13, I found that self-injury temporarily relieved the unbear-able jumble of feelings. I cut myself in the bathroom, where razor blades were handy and I could lock the door. The slicing through flesh never hurt, although it never even occurred to me that it should .... The blood brought an odd sense of well-being, of strength. It became all encompassing ....
> With a safe sense of detachment, I watched myself play with my own flow-ing blood. The fireball of tension was gone and I was calm ...

This can be contrasted with those beliefs, often based on religious funda-mentalism, that emphasise the person so much that the body is totally its servant, in some cases sacrosanct. The struggles that Jehovah's Witnesses have with the authorities because their beliefs allow no intervention into the body, or the dismay that many other fundamentalist religions have with post-mortem examinations, are founded on views of the relationship between the person and the body quite different from those that are the dominant ones in Western society. Belief that the body *is* the person leads to the view that any modification of it is a violation, just as in earlier days any amount of drawing and quartering was permitted because it could drive out the devil. Apparently similar beliefs led the Inquisition to assume that the truly insane were so dissociated from their bodies that they would not really feel pain.

A number of anthropologists throw further light on this interplay between the body and the person. Of particular relevance are Alfred Gell's explorations of tattooing in Polynesia in previous centuries. His conclusions have especial resonance for understanding the range of mutilations and violations of the body that can be found in present-day societies. He echoes Michel Foucault by stating that 'it is through the body, the way in which the body is deployed, displayed, and modified, that socially appropriate self-understandings are formed and reproduced' (*Wrapping in Images*, p. 3).

Gell takes our understanding of bodily modifications a stage further by elaborating the different functions of tattooing in Polynesia. To greatly simplify his argument, for brevity, he shows that in some societies, for example Samoa, it is the process of inflicting the tattoo that is paramount. It is permanent evidence of having undergone that process. As Gell puts it, 'Tattooing is the perfect vehicle for the bodily registration of commitment' (*ibid.*, p. 37). Here, the body was modified as a way of exerting control over the person. It shows the individual's position in society. The person is shown to be subjugated to the social structure. The process of subjugation is also an acceptance into that society.

Such a 'registration of commitment' requires a society in which there are clear structures and hierarchies. The meaning of the modifications of the body are a consequence of the social processes in which they are embedded. This can be seen in the contrasting role of tattoos in those Polynesian societies that are rather different from those on Samoa, for instance on the small islands of Mangareva where the society is less clearly structured. There, tattoos indicate a person's particular significance. Gell argues that for these more devolved and inherently competitive societies it is the mark of the tattoo itself that is crucial, rather than the evidence it gives of the process through which the recipient has passed. In these high-tension societies with enlarged social distances tattoos label social distinctiveness.

This more individuating role probably has more in common with the use of tattoos in our own culture. It is no accident that it is a particular age group who submit to tattoos: young people at a stage where they are forming their adult identities. They want their bodies permanently marked to make some statement about the person they are.

## The battle for the person

There is a powerful belief system rooted in the knowledge that we are more than our bodies, but it is constantly challenged by our need to cope with our experiences as mediated by our bodies. The challenge that this paradox raises, of being both body and person, is resolved in many different ways by different people. But the most common strategy is to hold on to a firm dualism that distinguishes these two different realms. On the one hand, there is the body with all its animal trappings, which shares all its major characteristics with every other human. In terms of scholarly debate this fosters studies in the natural sciences in which humans are virtually interchangeable with each other because their bodies are essentially identical. Indeed many of the properties of those bodies are so close to those of other animals that they can be studied interchangeably.

It is out of this perspective that there is the constant search for biological bases to many phenomena such as criminal activity, aggression or other acts of violence and violation. Genetic make-up, brain damage or hormonal influences are held up as the primary causes of violence, aggression or criminality in general. But this has similarities to the perspective of the rapist seeing his victim as merely a body to be used, or the serial killer who keeps body parts as souvenirs of his deeds. The body is taken as all that is significant.

Such a view ignores all those immaterial aspects of personhood that so enthral disciplines running the gamut from anthropology to psychology by way of linguistics and theology. Here is the second realm, the differences between people or the contexts they experience, which are a recurring source of debate. Those aspects of an individual that make them unique come to the fore in considering them as people. Aspects such as their creativity, morality, passion, potential, or their particular point in the flux of cultures that they illustrate, are recognised as transcending the bodily functions that support them.

Throughout history it seems to have been the case that the belief in personhood was protected by elaboration of the distinction between the individual and the body. The soul, psyche, personality, mind, character and many other aspects of the person have always been regarded as quite distinct from their corporeal existence. Yet the paradox that is fundamental to being human is that the significance of any human body is in how it expresses its supra-corporeal capabilities. The spirit can no more throw off its mortal coil than the clay of which we are made can be recognised as a being without evidence of its character as a person.

The struggle with this duality of mind and body is at the heart of most human endeavour. It is a struggle that aims constantly to re-create the fiction of personhood in defiance of the laws of nature, a fiction that casts its protagonists into opposing camps. Sin and evil are the products of the flesh that must be fought with the weapons of the inherently virtuous spirit. The profane is all that which relates us to our animal past, whether it be the subconscious urges of a Freudian *id* or the apparently more scientific but no less pessimistic claims of evolutionary bases for aggression and survival. The sacred is to be found in the purity of reason and the contemplative arts that are as far from bodily functions as possible.

But when these protagonists share the same virtual reality (as they do for everyone who has some hold on actuality) then there is the constant need to attempt to modify one or the other to make the person who houses them more acceptable. The modification may come from upholding the significance of the mind and spirit as targets for manipulation and refinement in an attempt to distance them as far as possible from their degrading companion; or the body is modified and in extreme conditions violated in order to make it more virtuous.

### The person as product

The quest for the body beautiful is an interesting development of the corporocentric perspective. Some of this may be a search for a better quality of internal life but much of the quest relates to the way a healthy body symbolises a good person. After all, there is still the temptation to blame people for their physical handicaps, as statements from such significant trend setters as an earlier manager of the English football team made clear. I think there are some gory parallels with the sorts of serial killer like Jeffrey Dahmer for whom his victims were clearly bodies to be modified and manipulated. He wanted to turn them into some sort of willing zombie for his own gratification. For whose gratification is the willing shaping of bodies by plastic surgery, or the other possibilities that genetic modification may allow? Often the determination to produce the perfect body seems to be an attempt to make the person apparently more spiritually pure. Yet this is always doomed by the paradox that the more we focus on the body the less able we are to allow those aspects of the person that capture history and character to break free. It is in the transaction between self and non-self, the dialectical relationships between mind and body, that humanity emerges. Too great an emphasis on one or the other leads to barbarity and degradation, whether it is promulgated by genetic scientists or serial killers.

The complexities of the relationship between the person and the body are at the heart of many important debates about the impact of current biomedical discoveries. These debates are often confusing because the mind is equated with the brain and the existence of a person is ignored. It is these confusions that unbalance debates as wide ranging as the possible inheritance of personality characteristics or the cloning of humans. The argument generated by any attempts to show that we are only what are bodies make us is so heated because each human being feels that his or her memories and intentions, feelings and character are ignored by the focus on the body devoid of the person. Claims that genetics can explain mental prowess, that mood is simply a product of our physiology, or that two identical humans can be created by a laboratory technician, and all the other proposals that challenge the view we each hold of our own rich and complex existence as individuals, are indeed threats to the fundamental, core constructs on which our minute-by-minute transactions with each other are founded.

### Beyond the body

The focus on involuntary body violations may imply that developments in biomedicine are all negative. But the opposite is often the case. The move away from state-sponsored violation has been reflected further in the changing attitudes of the medical profession towards how they may mutilate their patients. When the body was crudely understood so that dentistry and medicine had to be intensely invasive then it was difficult for practitioners to deal with their patients as people. The contrast with the changes to the body were too great, as were the implications that they had for the changes to the person, which the surgeon could not control. So mastectomies and hysterectomies were commonplace in contexts that would not now be acceptable. The advent of more refined drugs and keyhole surgery has helped medical practitioners to rediscover the person they are treating. They can afford now to relate to their patients as people and, indeed, they can recognise that it is the person who needs to be treated not just the body. This is what is at the heart of attempts to influence the lifestyles of patients.

It is interesting how this has produced radical changes in the issues that are considered relevant in medicine. Dignity and respect for the patient can take on new emphasis and even override decisions about what to do to patients' bodies. I was interested to learn from my own dentist just how much dental practice has changed. I had been aware that teeth were often removed *en masse* in the earlier part of this century as a preventative measure. But I had not realised how much this had been enshrined in dental dogma. Apparently it was still the case in the 1960s that dentists operated under the slogan 'extend for prevention, cut for immunity'. It was an alien idea that a person's view of themselves was related to their relationship to their teeth. But the pride people have in their teeth is clear from the queues for cosmetic orthodontic treatment. Once again the recognition that the person and the body have to be considered together has changed the way both are dealt with.

### Emergent complexity

The mistake that murderers and other violent criminals make is a sort of category error similar to that made by many of the more ardently reductionist biologists. Because a person is an inevitable correlate of a body, they assume that they will know everything there is to know about people by knowing everything there is to know about their bodies. Similarly the murderer in his limited view of the person who causes him frustration or anguish can only see the body that needs to be removed, or in extreme cases, sees no person at all but only a body to be examined or exploited.

Murderers ignore the ways in which personhood emerges as an entity that has its own forms of complexity that give it qualities which cannot be found in the body alone. These ways are derived from the history and anticipated future, the memories and social transactions, social representations and cultural experiences that give any particular person their unique characteristics.

There are crucial parallels here with the arguments Brian Goodwin, former Professor of Biology at the Open University, has made about the organism emerging as something more than the sum of its genetic make-up. As he puts it in his challenging book *How the Leopard Changed its Spots*, life has a rationality to it that 'makes it intelligible at a much deeper level than functional utility and historical accident' (*ibid.*, p. 105). He argues that much of this rationality can be found in the mathematical inevitability of complex processes having properties that cannot be simply derived from knowledge of their arithmetically primitive constituents. For example, he cogently explains how the cell is much more than its genetic make-up and transmits much more to future generations that just DNA sequences.

What I have been illustrating is how one of the emergent properties of the complex systems that are human beings is the person. Any attempt to reduce this creation to bodily components will lose the entity it is describing. The body is one of the resources from which the person is made, just as genes are resources that help to create cells.

Functional utility and accident also do not go very far to explain the significance of the voluntary and involuntary violations of the body. They are embedded in a psychosocial process that gives significance both to the body as object and to its reflection of the body as subject, the person. By ignoring the significance of the person and focusing on the body, violent criminals teach us the civilising influence of recognising the importance of the person. This is a lesson that many scientists seeking to help humanity rather than destroy it would do well to master.

FURTHER READING

Canter, D., *Criminal Shadows*, London: HarperCollins, 1995.

Chandler, R., *The Long Good-Bye*, Harmondsworth: Penguin, 1951.

Gell, A., *Wrapping in Images*, Oxford: Berg, 1993.

Goodwin, B., *How the Leopard Changed its Spots: The Evolution of Complexity*, London: Phoenix, 1994.

Innes, B., *The History of Torture*, Leicester: Blitz Editions, 1999.

McAdams, D.P., *Power, Initimacy, and the Life Story: Personological Inquiries into Identity*, New York: Guilford Press, 1988.

# 5  The Dead Body and Human Rights

THOMAS W. LAQUEUR

### Introduction

As any reader of detective fiction will know, the *corpus delicti* is 'the body of the crime – the body, the "material substance" upon which a crime has been committed, for example the corpse of a murdered man, the charred remains of a house burned down. In a derivative sense the substance or foundation of a crime: the substantial fact that a crime as been committed.' 'The corpus delicti is the body or substance of a crime which ordinarily includes two elements: the act and the criminal agency of the act.' These definitions are laid out by Henry Campbell Black in his *Law Dictionary* (1990).

The establishment – more precisely the unearthing – of these 'substantial facts' and their identification has become a central and much publicized aspect of international human rights work. Since 1985, the Mothers of the Plaza de Mayo have been conducting a successful campaign to name at least some of the skeletal remains of the disappeared 'No Names'. Under a picture of trousers and shoes and other articles of clothing laid out on the floor and labeled with the names of their former owners, the *New York Times* assures us that forensic pathologists are examining the bodies of recently murdered Ratchak villagers. And we learn that the Serbian authorities did what they could to hinder the embarrassing autopsies which would make clear that, no, these bodies did not die in the course of ordinary military action but were then set up by the Kossova Liberation Army (KLA) as if they had been massacred. A policeman of a South African apartheid era hit squad confesses to an array of abductions and murders but this is not truth enough, either for the authorities or for the parents of the dead. He leads a team of investigators to an abandoned farm where forensic pathologists unearth the body of a woman buried in a crouching position, a neat hole where the bullet entered her skull, a more ragged break in her jaw bone where it exited. The bones are taken to a morgue and there the parents make an identification – the high cheekbones are the telltale sign. The policeman's confession fills in the details between the name of a once-living girl in her 20s and the skeletal remains. And so too in Vukovar and Srebrenica and in Rwanda.

My question in this essay is why the *corpus delicti* has assumed this prominence in how we think of human rights violations and whether this attention to the particular, corporeal 'substantial fact that a crime has been committed' makes any difference. Whether, as the historian Carlo Ginzburg put it in another context, carefully interrogating each suspicious body might restrain 'our power to pollute and destroy the past, the present, and the future'.

### Who is the *corpus delicti*?

The – or a – *corpus delicti* stands at ground zero, the very beginning, of a narrative of truth and specificity. A person, or of course much much more likely in our century scores, hundreds, thousands, tens and hundreds and thousands of thousands of persons, did not just disappear. We, or more precisely their Polish countrymen, were not satisfied with the explanation proffered by the Soviet Union in 1941 that 15 000 officers and men were missing because by some miracle they had marched through a war zone and across Asia to Manchuria; nor were the Argentine mothers content to believe that their missing children were on holiday in Europe as the military regime once claimed. The Jews of each and every east European *shtetl* and town are not simply gone as by some impenetrable, occluded act of nature. But even, as in the former Yugoslavia, where there was little if any attempt to hide the crime, and where insufficient time has passed to naturalize absence, the recently unearthed bodies reveal a truth not evident from satellite photographs or even onsite inspections. This and that man is gone; he is dead; his body reveals he did not die from natural causes; there is a bullet wound in the head. Someone caused it to be there. And, we (forensic experts, lawyers, an international public) can from this kernel of truth – this 'body or substance of a crime which ordinarily includes two elements: the act and the criminal agency of the act' – write a narrative with political, juridical, and more intimate memorial, therapeutic consequences.

The name of the body may or may not be legally relevant – the murder of one or many unknown people is still murder, the genocide of a people, however anonymous to the outside world is still genocide – but it is emotionally, hence rhetorically, and hence politically exigent. The murdered woman, man, or child is not a cipher, the genus in question is not the beasts of the field or the beasts with which we live in great intimacy – dogs, as in 'to die like a dog' – but the human genus, which names its members and makes these names a central element of identity, memory and commemoration.

In a way all of this may seem obvious, but the body-by-body, name-by-name, standard of truth is absent in reports in other ages of what we today would call human rights abuses. Thucydides' account of how the Syracusans starved to death 7000 prisoners of war goes by in less than a paragraph of his *History of the Peleponesian War*. 'They were', we are told 'crowded in a narrow hole, [now on the itinerary of the historically minded tourist in Sicily] without any roof to cover, the heat of the sun and the stifling closeness of the air tormented them during the day – each man during eight months having only half a pint of water and a pint of corn given him daily.' And Old Testament accounts of slaughter are even briefer and less informative. To be sure, the perpetrators were following higher orders and were elsewhere enjoined to mourn in some fashion the victims of God's wrath. (I am thinking of His reproof of the pleasure the Israelites took at the destruction of their Egyptian captors.) But that said, orders are given and manifest savagery is recounted in bursts scarcely a sentence long: 'you shall save alive nothing that breathes' (Deuteronomy 16:21); 'Then they destroyed all the city, both men and women, young and old, oxen, sheep, and asses with the edge of the sword' (Josiah 6:21); 'So they slew him, and his sons, and all his people [the Amorites] until there was not one survivor left to him' (Numbers 21:35).

### Human rights – war crimes

Technically, of course, the *corpus* became a *corpus delicti*, an articulate witness to a crime in the context of human rights, only when crimes against humanity or genocide became crimes: 20 December 1945 Control Council Law no. 10, Punishment of Persons Guilty of War Crimes, Crimes Against Peace, and

Against Humanity, or 10 December 1948 the Universal Declaration of Human Rights. But the individual body as the substantial fact of a crime and of criminal intent has deeper roots even if it does not figure prominently in Thucydides or the Old Testament.

In some measure it is a response to a problem, first raised by Kant, for the aesthetic imagination that translates, I would suggest, to the moral imagination as well. It is the problem of the mathematical sublime. The arithmetician has no more difficulty in principle in comprehending one murder than 600000 – the number murdered in the Armenian genocide of 1916–1917 or by Nazi Einsatzgruppen on the Eastern Front in 1941 before the death camps were fully geared up – or 5000000 to 6000000, the best estimates we have of the number of Jews murdered in the Shoah. At a purely cognitive level any number can be understood by adding, unit by unit, to the unit that comes before. But, Kant says, actually being able to take in great magnitudes – to feel their sublime terror – is ultimately an aesthetic act that depends on gaining the right distance on the subject. His example arises from the case of the Great Pyramid. Too close and we see only stone by stone and cannot take in the full sweep from base to peak; too far away, we lose the perception that this massive structure is built stone by stone, and we stand to lose the sublime wonder that arises from apprehending that something so massive was made block by discrete block.

The substantial fact of genocide is the accumulation of the substantial facts, particular by particular, the corpus delicti, one by one, each manifesting 'the act and the criminal agency of the act'. It is another question whether we can best comprehend its sublimity by an intense, almost synecdochic, attention to the fate of one body (that of Tomislav Levic, exhumed from a pit on the Ovčara farm in Bosnia), or through some temporal register (more than three people murdered per minute, 4000 every day for four months in Rwanda), or in some massive whole (6000000), which we must then force ourselves to imagine again as created one by one. But transforming a body into the 'substantial fact of a crime' is a necessary if not sufficient comprehending of the terror of mass murder: some distance will allow us to see that the pyramid is built of individual stones.

### Disbelief

The body as witness has come to prominence also because of another relatively modern problem: incredulity, disbelief, *ennui*. When Milton Bracker,
an American reporter, toured the abandoned Natzwiler concentration camp
in 1944 he says that he had constantly to remind himself that, in this beautiful
setting, the site was not a forester's camp but a charnel house. He saw the
crematoria, hooks for torture. He knew, he said, they were not 'used for mere
decoration'. And yet he writes in the skeptical voice: 'It is reported that . . .'
He could not vouch for the deaths of 16 000. Famously, Felix Frankfurter
discounted even eyewitness evidence of mass extermination. He was horrified by the stories told him by Jan Karski but he did not believe them. No,
he clarified his position: he had no doubt Karski was telling the truth; he
simply could not believe him. It was all too easy to be on guard against being
fooled by these new atrocity stories as so many had been taken in by the
propaganda of the Great War. One could submerge five or six million Jews
into the even larger mass of twenty-five million dead bodies left in wartime
Europe: dead from starvation, military action, cold, heat, natural causes come
prematurely. The simply dead.

### Evidence

It was the pictures of stacked bodies, strewn corpses – the photograph as
dissection – that at least to some extent finally produced the sense of a substantial fact both in the juridical context of the postwar trials and amongst
the general public. These were indeed *corpora delicti*. Balzac suggests why
this might be so. 'All physical bodies', the photographer Nadar reports him
as saying, 'are made up entirely of layers of ghost like images, an infinite
number of leaf like skins, laid one on top of the other'. When someone was
photographed, one such spectral layer – one thin film of the essence of life
– was transferred to the print. Something of this strange image – poetically
suggestive and at the same time materialist, hard-edged, anatomical – may
explain why in the absence of the body the photograph bears such weight
in modern human rights reporting.

Finally, the dead body has come under increased scrutiny as a *corpus delicti*
because, more generally over the past two centuries, whole categories of death

have been singled out as more than ordinary, to be expected, routine. The history of death intersects with the much more general history of causality, of how the lifeless or injured body came to be that way as the consequence of a particular step-by-step, agent-by-agent, cause-by-cause, sequence of events and not simply as an Act of God, as a part of the order of things, an unfortunate consequence of forces beyond human ken or control. In one sense this is the story of the spectacular modern rise of the autopsy. Giovanni Morgani praises his seventeenth century predecessor Théophile Bonet for publishing as great a number as possible 'and digesting into order the dissections of bodies', thereby forming them into 'one compact body'. Useless, random information thus became something extremely useful. All known autopsies in 1679 fit into one volume. By the early nineteenth century they would fill a small library.

### Death as a social disease

An explosion in post-mortem investigative dissection, however, did not alone enable the modern human rights inquest. Death, as nineteenth century epidemiologists increasingly discovered, is a social disease and society itself is not a fact of nature. Miners did not die in an explosion because of the unfortunate natural existence of methane deep under ground but because mine owners failed to provide proper ventilation or proper fire doors. Slaves did not die during the middle passage because of the unavoidable rigors of sea travel, nor did they suffer as a consequence of a natural condition of slavery. Specific actions carried out – or omitted – by specific people caused death and injury. Thus, beginning in the eighteenth century, morbid anatomy and organic pathology would become the foundation not only of forensic medicine and anthropology but of a humanitarian narrative more generally, of case-by-case narratives in which someone acts, or can be shown not to act, in such a way as to cause harm to other human beings each with their own subjectivities and rights. It is this chain of causal specificity that demands either compassion or legal intervention unlike, for example, the general accounts of slaughter in Thucydides or the Old Testament, which sustain little more than revenge.

I should perhaps pause for moment to declare an interest. My father was a pathologist who worked both as a medical examiner in the USA – a coroner in the English sense – and as a physician to miners in a union hospital. He was actively involved in a long medico-political campaign, as was I in a different capacity, to make a disease called black lung (pneumoconiosis) compensable in West Virginia and in federal law as it has for a long time been in the United Kingdom. The critical issue was whether the manifest breathing impairment of long-time workers at the coal face was due to their exposure to coal dust, as one might assume, or to some extraneous cause to which anyone might be subject. As the law then had it, if X-rays revealed hard opaque lesions in the lungs then the miners' specific work-related exposure was judged to be the cause of their suffering. But if, as was often the case, long-time miners who could not walk up three stairs without much huffing and wheezing did not manifest these shadows, then their impairment was taken to be part of the general order of things. They had smoked or they simply had bronchitis or emphysema just as anyone else might have. No *lesion delicti* in other words.

My father and his colleagues showed by doing autopsies on these sick miners when they died that their lungs had microscopic lesions that were largely absent in miners without breathing impairment and absent entirely from the lungs of the general population. Indeed they were absent in the lungs of Welsh miners who had worked in pits where levels of coal dust had been much reduced. Industrial health may not be among internationally recognized human rights but since the pioneering research of Bernardino Ramazzini in the eighteenth century it is a domain in which the truth about work is found in the body. It is of course not a truth that speaks for itself but it does create limits to interpretation. A 'substantial fact' will not go away.

## Truth

In contemporary human rights there are two related, sometimes divergent, sometimes parallel, uses of the truth of the *corpus delicti*: broadly speaking, truth in the interest of political advantage or prosecution (justice) and truth in the interests of memory, of narrative closure, of healing, of reconciliation. The first depends upon the fact that the body is indeed the body of a crime and not simply a dead body. The dirt is stripped away, extraneous details put to one side and the circumstances of death established in sufficient detail to point to a perpetrator. Identification is not necessary. The second depends crucially on naming. A person, with an identity and a place within a family and a community, is brought back from the anonymity of the grave and in some sense given back to his or her people. The rhetoric of justice is manifestly different from the rhetoric of conciliation. I will come back at the end to the question of whether the one is necessarily a substitute for the other.

## How the Nazis invented human rights inquiry

Ironically – almost a joke – the first major forensic inquiry to establish human rights abuse was undertaken by the Nazis in an effort to demonstrate that the Soviet Union had carried out egregious violations of the 'laws of war' against the soldiers of its erstwhile ally, Poland. The purpose of this extensive inquiry and of the document it generated – *Official Material on the Mass Murder at Katyn* – was, and remained in future incarnations, glaringly political. The Germans wanted to embarrass the Soviet Union and sow disunity amongst the allies; the Soviets immediately sought to discredit the report and after the war imprisoned two of the pathologists to make them change their testimony so that the Germans could be charged with the crime at Nuremberg. In 1952 the US Congress elaborated the original German material in a massive 1800 page *Report of Hearings Before a Select Committee to Conduct an Investigation of the Facts, Evidence, and Circumstances of the Katyn Forest Massacre*, which appeared as a salvo in the McCarthy era cold war against the Soviets. That said, an international team of pathologists whose integrity has not been seriously questioned exposed the *corpora delicti* of a massive violation of the human rights of prisoners.

The background story briefly is as follows: 180 000 Polish troops were taken captive by the Soviets in 1939. In July 1941 when inquiries became possible it appeared that some 15 000 of these, including 8300 officers, had disappeared without trace. Letters home had abruptly ceased. For almost two years Polish authorities, through a variety of channels – there were fifty formal requests for information – demanded some sort of accounting but got nothing in return. The men had escaped, Stalin told General Sikorski in December 1941. Where to? To Manchuria. Yes, across Asia to Manchuria.

Soon they would be found. Early in February 1943 a soldier of a German army unit bivouacked on the site of a former Communist Secret Police (NKVD) villa near Smolensk digs a latrine trench and comes upon a mass grave. (The Germans, of course, had by this time invaded Russia and the parts of Poland that the Russians had occupied in 1939.) The bodies are those of Polish officers; the bullets with which they were murdered were of German manufacture, a potentially embarrassing fact; but this was one atrocity which the German Ministry of Information soon confirmed its forces had not committed. On 13 February, Goebbels announced that the disappeared Polish officers – at least some of them – had been found, dead at the hands not of the Poles' enemies but of their ally, the Soviet Union. On 15 April, the Soviets replied that, yes, these bodies were indeed those of the disappeared Poles, but, no, the blame was misplaced. The men were prisoners of war whom the Germans had captured when they conquered the area in September 1941. This response, at least, was more serious than that the disappeared had escaped to Manchuria and a great propaganda flurry ensued.

In response to the Soviet counterclaim, Goebbels organized the largest human rights inquiry ever, an investigation that had its roots in nineteenth century German forensic medicine and one whose basic form would echo into the present. International experts were summoned, the Red Cross was called in to witness, field laboratories were set up, bodies were exhumed, locals interviewed, photographs taken, clothes and other material from the mass graves carefully scrutinized, moldering identity cards removed and matched to lists of the missing, the narrative of death built up detail by detail. At the end of the day, there was the *corpus delicti* or, to be more specific, some 4300 *corpora delicti*. (About 11 700 prisoners from two other camps have never been accounted for: missing, still in captivity.)

Much of the Katyn investigation would seem painfully familiar to the forensic pathologists and anthropologists who have worked since in the Argentine, or the former Yugoslavia, or South Africa or Rwanda. It was about establishing one of the two critical aspects included in the *corpus delicti* – 'the criminal agency of the act'. (The fact of the crime was clear enough.) Thus it was shown that the rope with which the prisoners were bound had been cut into uniform lengths, suggesting careful preparation; microscopic analysis of its fibers proved that it was of Russian manufacture. The uniform, regular pattern of the skull wounds – at the base so as to hit the medulla, powder marks at the point of entry – suggested that the prisoners had been restrained and systematically, knowledgeably shot. Random bayonet pierces in the coats of the exhumed suggested that many had struggled before being bound and held; examination of the cloth tears revealed that they had been made with a four-sided sharp instrument, the shape of the Soviet – but not the German – bayonet.

But the weight of nineteenth century forensic science and the idea that the time, cause, and manner of every death could and should be accounted for is most evident in the intensity with which the bodies were examined to determine when exactly they had died, or more specifically when they had been buried. Nine detailed autopsies were performed on previously unexhumed bodies – the *de novo* exhumations themselves were carefully witnessed and photographed; 982 further bodies were examined; 4143 were accounted for. (Note the numerical precision that will come to characterize not only human rights exhumations but the memorializing in our century of mass death more generally.) The question was whether the men represented by these nine bodies had been murdered in the late summer or autumn of 1941, when the German's captured the area, or earlier at the hands of the Soviets. On whose watch had they 'disappeared'?

And so Dr Ferenc Orsos, professor of judicial medicine at Budapest University reports that on the basis of his experience – some 80 000 autopsies, numerous scientific papers – the brain pulp was so calcified as to prove that the corpses had been dead for about three years. His distinguished colleagues agreed. Dr Edward Lucas Miloslavich, American trained, for a time professor of medico-legal pathology at Marquette University, and after 1934 at the University of Zagreb, presented evidence on the basis of his vast experience that the muscles of the exhumed bodies, taken from various body parts, showed upon microscopic examination no signs of striation and a degree of saponification incompatible with burial of less than two and a half years. And so on, pushing the date of death well into 1940 when the area was unquestionably under Soviet control.

### Accounting for the dead

The medical expertise that made the Katyn Forest investigation – and modern human rights investigations more generally – possible grew out of two allied earlier developments: the late eighteenth and nineteenth centuries state's interest in having every citizen accounted for and, more specifically, every death duly registered and classified and also doctors' interest in expanding the purview of their science from individual to society. The political views of the German fathers of forensic science varied greatly – Rudolph Virchow was a liberal and passionate supporter of the 1848 revolution; Ludwig Caspar, on the other hand, inveighs against the rioters and declares that of eleven deaths only that of a city militiaman was 'honorable' – but they shared a passion for making the dead body bear witness to its history. Virchow essentially invented the modern autopsy protocol, the careful step-by-step, orderly examination of a body and its surroundings; the accumulation of mountains of reliable data correlating diseased organs, broken bones, bullet wounds with clinical and other evidence. Caspar, along with French colleagues, developed this method into our modern practice. Do not be deceived by the open knife, supposedly covered with blood, found near a much-decayed body, even if a knife of such poor quality would probably not belong to so well-dressed a corpse. Maybe murder. But no. Despite the putres-cent state of the body, the expert can see that the right side of the heart and the pulmonary artery are distended. Apoplexy of the heart. And in any case, there is no knife wound. Just a body, not the body of a crime. And so on through hundreds of exemplary corpses.

The technical expertise to discover the truth about bodies and one of the continuing uses of this truth is thus juridical and more broadly political. Again, ironically, it is because even the criminal generals who governed the Argentine could not quite abandon the nineteenth century prescription that every body had to be accounted for that the *corpus delecti*, the substance of their crimes, was discovered. Some bodies were thrown out of airplanes or simply dumped during the dirty war, but most went to the morgue where, following nineteenth century laws regarding the dead, they were duly regis-tered – sex, age, often cause of death, place of burial, time of burial. On the basis of this record generated by the state it could be shown, for example, that the proportion of nameless people dying from gunshot wounds increased from just over 5% before the coup to over 50% two years later. It was this trove of information that allowed the teams of American and Argentine forensic pathologists and anthropologists to identify the bodies of hundreds of 'the disappeared', the officially designated 'no-names' (N.N.), and to reveal, in the mass graves of public cemeteries, the existence of thousands more.

In a way then, the Argentine project – or the more recent Rwandan forensic investigation – is not so different from the German project: the young people were not in Europe; the heaps of Tutsi bodies did not belong to the victims of an unfortunate civil war but had been hacked to death at close range – every man, woman, and child – by machetes; the Polish prisoners had not escaped to Manchuria nor had they been recently shot by the Nazis. The exhumations in Vukovar and Srbrenica – and in the cemeteries of various Argentine cities – certainly provided substantive facts of the crime, and the scientists working on them were unquestionably motivated at least in part by a desire that justice be done; in other words, they were seeking the truth for juridical purpose. But there was in Argentina and also in the Bosnian and South African cases something else going on as well. The *corpora delicti* had another function as well in situations where the fact of a crime was already well enough established.

### A need to mourn

Serbian forces made no attempt to hide their deeds, even if some people made themselves believe that the missing prisoners were alive somewhere, working in the mines. There is no secret to be revealed, even if it is important that eyewitness reports of mass murder be coordinated with mass graves of blindfolded and bound bodies. The body, is, to be sure, a *corpus delicti* but it is also proved to be the body of someone in particular. Its identity is probed and its fatal history is written, less for what all this says about the perpetrators than for what it says to survivors or rather, for what it allows them to say and feel about each individual in the mass grave. A passion to remember, a need to mourn and to heal a psychic trauma and a social wound seem to demand that each, and every, body be accounted for by name. The mass grave – the 'pyramid' of death to use Kant's metaphor for the ungraspable vastness of the arithmetic sublime – is disarticulated into its singular components through as detailed a reconstruction of the individual narrative of death as possible. The *corpus delecti* speaks now less of crime and intent but of the finality of death – the *dénouement* of a life. The power of the magnitude gives way for the moment to the intimacy of the singular, the detail.

'Without bodies and funerals', Eric Stover writes, many of the Muslim women of Bosnia 'could not visualize the death of their husbands and sons and thus accept it as real'. In the absence of a body, there was the hope against hope – a 'survival mechanism' to use that odd modern therapeutic term – that maybe, maybe the men were alive somewhere. This strategy of denial had clearly run its course. Closure – which meant finding and naming the body – seemed the only emotionally possible beginning for a survivor's new life, even when official death certificates had already made one legally possible. (This need for a closure is itself a claim worth investigating. My grandmother for over twenty years seemed to find great comfort in the idea that her daughter was quite possibly lost somewhere in Poland – or more vaguely 'the east' – and had no interest in hearing the account from a niece who saw her die *en route* to a concentration camp.)

Physicians for Human Rights and other forensic teams produced definitive death histories in the former Yugoslavia. Tomislav Levic, for example, was a pharmacist who left Zagreb for Vukovar to work as a medical aid; he was wounded and taken to a hospital there. The hospital was taken by Serbian forces, the patients – military and civilian – were taken to a nearby farm and murdered. A burial pit was identified, in some part because catheters and saline tubes remained attached to the decaying bodies, as their mass grave. One of the first bodies exhumed bore evidence of an old arm fracture; Dr Strinovic called Lesic, Tomislav's wife, and asked if she had any medical records. Yes, she had an X-ray of an old arm fracture and a dentist still had her husband's records as well. He was identified, although this fact was not made public until a ceremony honoring all the dead could be organized. Mrs Levic, however, allowed herself to light a candle during a visit to Bethlehem in memory of her found, dead, husband and another candle in honor of her son's father. The son stood proud at the public funeral when the name Tomislav Levic was read out. The widow took the keys that were found in the corpses' pockets and, after the funeral, used them to open her apartment door. The *corpus delecti* has become the dead body of a specific, named man who is recouped for his kin and perhaps for his community. Justice, at best, waits in the wings.

In the Introduction to this chapter, I mentioned a case that Michael Ignatieff has recounted from South Africa, in which justice is even less relevant. The perpetrators had confessed and were given amnesty in return. So, establishing the fact that the victim had been tortured in a effort to make her turn informer and then, when that failed, stripped, bound and shot, as the body itself had testified, is purely for the benefit of the parents, who found solace in knowing the truth about how their daughter died: honorably. Strangely each excruciating detail of her suffering seems to have been a comfort to her father.

### Naming the dead

This naming of the dead from violence, telling their story, recouping them for the world of kin and progeny, has its roots in antiquity. One thinks of the lists of Athens' fallen soldiers – the so-called Patrios Nomos – and of course of Pericles funeral oration as an act constituitive of the polity of the living and the dead. (Pericles of course did not actually recite the names but they were before him on tablets.) That said, the names of ordinary people are largely lost until relatively recent times and great armies, whole villages, disappeared with nary a nominative trace.

> Then shall our names,
> Familiar in his mouth as household words –
> Harry the King, Bedford and Exeter,
> Warwick and Talbot, Salisbury and Gloucester –
> Be in their flowing cups freshly remembered.

Says King Henry V (Act IV, Sc. 3)

> This story shall the good man teach his son,
> And Crispin Crispian shall ne'er go by
> From this day to the ending of the world,
> But we in it shall be rememberèd,
> We few, we happy few, we band of brothers.
> For he today that sheds his blood with me
> Shall be my brother; be he ne'er so vile,
> This day shall gentle his condition.

In fact, very little is offered the common man in this magnificent prolepsis of immortal fame. Poetry full of names, naming and the promise to name

actually names only those of noble blood and a saint. Perhaps he that be
'ne'er so vile' can have his condition gentled, i.e. can get a name, can be made
part of a community of blood; but only through some secular shadow of the
eucharistic miracle are the vile and un-named gentled and specified and made
'a brother'.

The betrayal of memory of which this speech only whispers will become
more resonant in the herald's famous accounting of the human cost of battle
some moments later. 'Where is the number of our English dead?' Henry asks
after what would become known as the Battle of Agincourt: he is handed a
piece of paper and reads

> Edward the Duke of York, the Earl of Suffolk,
> Sir Richard Keighley, Davy Gam esquire:
> None else of name; and of all other men
> But five and twenty . . .

The 'band of brothers' turns out to be those few named. The remainder
– 'none else of name' – who were promised immortality but a few scenes
earlier, are absent from the theater of history and from memory, at least
from public memory.

Literal namelessness is, of course, not what is at stake here, although
Shakespeare's reduction of the number from 500 to 600 given in his source
emphasizes how very exclusive was the pool for the band of brothers. We
do, in fact, know the names of all persons of whatever rank who served at
Agincourt because Sir Robert Bapthorpe, the Controller of the King's
Household, prepared a list of them. From this list and from indentures and
other documents we know, and Shakespeare's audience could have known,
that the names of the four archers in the service of the felled Sir Richard
d'Ketly (or Keighley), for example, were William de Holland, John Greenleaf,
Robert d' Bradshaw, and Gilbert Howson. They and the common men in
*Henry V* did, of course have names, if generic ones; they do in all cultures –
Pistol, a nickname, or Michael Williams, the stock Welshman. But these are
not what Shakespeare means by 'name' in 'none else of name'. Name in *Henry
V* and other Shakespearean contexts is a sign of blood, of lineage, of patriarchy, of place, of kinship in the profoundly cultural sense of a band of broth-
ers somehow subsumed in the name of the king.

In those days it might be worth the effort to identify the mutilated body of a Charles the Bold – much ingenuity actually did go into the process – but then he was the Duke of Burgundy. Whole villages were torched and their inhabitants anonymously killed or scattered in Bohemia and elsewhere during the Thirty Years War. Not so in our century. We have the names of everyone whom the Germans executed in the destroyed Czech village of Lidece; scores of memory books seek to reconstruct with names and pictures the lost souls of Europe's Jewish communities.

How more and more of us came to have names and identities that matter – names that are worth recording, remembering in public and in private, is another story but it is one of the features of nineteenth century societies that increasingly 'none else of name' would no longer do. In fact tens of millions of names came to stand as memorials, first in the hundreds of nineteenth century memorial parks that followed the founding of Père Lachaise in Paris in 1804, then on the battlefields of the American Civil War, and most dramatically on the millions on gravestones and vast commemorative walls of the Great War. Maya Lin's Vietnam memorial of names was, of course, born at Thiepval in the Somme. In 1918, too, was invented a memorial to the poignancy of namelessness – the 'Unknown Warrior', *le soldat inconnu*, a phrase translated over and over again in our century: in 1943, in the Warsaw Ghetto for example, Chaim Kaplan would speak of the day 'when the Jewish people will erect a memorial here, where the common grave holds all these brothers forever. Here lie our "unknown soldiers", whom all of us should honor and remember.' (Sisters are not in this band of brothers.) Subsequently, unprecedented numbers of names that in earlier generations would have been lost now mark the contours of more battles; they are attached, or point, to the sites of slaughter and betrayal, the pauper graves and burial pits, the ash heaps in concentration, the sick beds of over 70 000 dead from AIDS.

No bodies were more irretrievably lost than those of the millions who perished in the Nazi gas ovens and perhaps in no other context has more effort been expended to restore a *corpus delicti*. These *corpora delicti* do, of course, bear judicial witness to a crime but in this respect their voices are moot. Punishment is no longer at issue. We name whom we can so as to extract a person from a multitude – from the ashes – and to remember, mourn, or simply contemplate him or her. I am thinking not only of the walls or trees of names but of the books and rooms of photographs – the spectral layers of a person – that have been so painstakingly compiled.

### Picturing the dead

Consider as my last example of the body in human rights the work of Serge Klarsfeld. He has compiled a master list of 75 721 names, all known deportees from France. This he winnowed to a more limited category, a list of 11 400 boys and girls under eighteen years to which he added the addresses from which the children were taken and the places to which they went. A name can be placed in some fixed set of coordinates in France and then traced to the gas chambers: Daniel Brunschwig, shown in two pictures, aged 3 or 4, standing in a garden next to his seated mother in one, his dad in the other, was taken from 28 rue du Titien, Cannes, and from there to an assembly point in Nice, and from there to Drancy and from there to Auschwitz on Convoy 61.

Week by week, place by place this happened. We know where the bodies started. And 2502 photographs – all that he and a team could find of the total 11 400 children – work like the exhumed bodies from Bosnia or Buenos Aires. They too were people. Some pictures are beautiful, some charming, some pleasingly conventional, some technically incompetent. In some the children look happy and fetching, in others tolerant of the occasion. Some look dumb as kids are wont to look when adults for reasons of their own insist on taking a picture. In short, what one would expect from any collection of pictures except for the jarring and very public Star of David, which feels eerily as if it had invaded the private space of the pictures without its bearers having noticed. Some pictures Klarsfled has arranged alphabetically but some not. It is not clear why Georges Lyon, shown sitting on a satin-covered box and looking as if he would rather be elsewhere than having his picture taken in what must be a 1928 or 1929 studio picture, is next to Elisa Zytaner of whom we know only her birth date, that she was arrested in the Toulouse area, and deported on Convoy 77. The project clearly proceeded episodically; we do not know why one body in a pit is next to another either. The overall effect is once again to insist that the sublime of mass death is produced from the arithmetic procedure of addition, one child and then another child.

### The tension between truths

There is clearly a bridge between the two functions of the dead body which I have recounted in this essay: the story of how each body became a *corpus delicti* leading to prosecution or some sort of political action and the story of how each body becomes the site of mourning, remembering, remaking of self and community. The link of course is that these are not the bodies of the beasts; they did not 'die like dogs' outside of law and culture. Or rather, they did 'die like dogs' despite the fact that they were human – that they were individuals, that they were conscious of their state, that they lived in communities – which is why it is so important subsequently to determine their identities and their histories. But that said, there is also a tension between, on the one hand, truth for the purposes of remembering, more broadly truth as some sort of individual communal therapy, and, on the other hand, medico-juridical truth, which grounds legal or political action. The rhetoric of memory is manifestly different from the rhetoric of justice; the question is whether the one might serve as a substitute, an excuse, for not pursuing the other.

In a sense I am only mapping the well-established tension between truth and reconciliation on the one hand and justice, or even vengeance, on the other. The former may have to be bought at the price of the latter. Or I may be mapping the even more venerable tension so evident in twentieth century Marxist opposition to psychoanalysis. The point is not to reconcile oneself to the world but to produce changes in it.

I do not want to claim that there has been a disingenuous trade-off on the part of the West in many of the recent human rights inquiries. Those who help to identify bodies and discover how they died are not the same people who refused to help when something might have been done to prevent the murders. But there is a deeper tension. The will to prosecute – to reconstitute a community juridically – may well be blunted by whatever peace remembering brings. It is not clear that the named bodies of the dead will serve us as both a *corpus delicti* – 'a body or substance of a crime which ordinarily includes two elements: the act and the criminal agency of the act' – and as the balm of closure. Putting them to rest may, though one hopes need not, mean putting much else to rest as well.

FURTHER READING

Joyce, C. and Stover, E., *Witnesses from the Grave: The Stories Bones Tell*
Boston, MA: Little, Brown and Co., 1991.

Laqueur, T., 'Bodies, details, and the humanitarian narrative', in *The New Cultural History*, ed. L. Hunt, pp. 176–204, Berkeley and Los Angeles: University of California Press, 1989.

Laqueur, T., 'Memory and naming in the Great War', in *Memory and Commemoration*, ed. J. Gillis, pp. 150–167, Princeton: Princeton University Press, 1993.

Thorwald, J., *The Century of the Detective*, New York: Harcourt, Brace, 1965.

US House of Representatives. Select Committee on the Katyn Forest Massacre, 82nd Congress, 1st and 2nd Sessions, 1951–1952, 7 parts. Washington, DC: US Government Printing Office, 1952.

Stover, E. and Giles, P., *The Graves: Srbrenica and Vukovar*, Berlin, New York: Scalo Books, 1998.

Zawodny, J. K., *Death in the Forest: The Story of the Katyn Forest Massacre*, New York: Hippocrene Books, 1988.

# 6 Nude Bodies: Displacing the Boundaries between Art and Pornography

GRISELDA POLLOCK

## Introduction

Let me start with a shower scene (Figure 1). Not Hitchcock's famous encoding of naked feminine vulnerability and sadistic masculine voyeurism mingling sex and violence that culminated in a hysterical act of surrogate matricide in his film *Psycho* (1962), my example is a work of video art by the contemporary Palestinian artist working in London Mona Hatoum (b. 1952). Titled *Measures of Distance,* made in 1988, the video is composed of several sound tracks working with and over several layers of imagery to convey the pain of separation and the longing of exile as well as the affirmation of a culturally specific feminine subjectivity and sexuality. Across what is at first only a densely pixilated field, grainily infused with intense colours forming at first indecipherable forms, lies a superimposed grid, a blow-up of lined but now transparent writing paper covered with a handwritten Arabic script. On the sound track, as if in an interior, there is the continuous sound of women's voices, chatting in Arabic and occasionally laughing. Two forms of intimacy are layered onto each other. From time to time, this ambient sound is overlaid by the carefully modulated tones of Arabic-inflected English in which a woman reads the loving letters addressed to an absent daughter from her mother. The speaking woman is the artist Mona Hatoum, working in London and making the video in Vancouver. Her mother lives in and writes from the Palestinian refugee settlement of Shittilah, in Beirut, which the daughter artist revisited during the mid-1980s.

The slowly changing video image is composed of photographs screened on a grainy surface, suffused with colour that only slowly renders up the image of a mature woman, naked, taking a shower. We must read the imaged body as that of the mother whose letters form a screen before it, and whose

FIGURE 1. *Measures of Distance*, detail from a video by Mona Hatoum, 1988.

thoughts and feelings are enunciated across the image by the voice of her now distant and absent daughter. In an interview in 1996 the artist commented on the relation of this project to the recurrent anxiety so often expressed in feminist art theory about any representation of the naked female body, given the profound appropriation of that image by a pornographic, voyeuristic, or sexually exploitative gaze that underpins most of Western visual culture.

> Well, in early feminism the attitude was that any way of representing a woman's body is exploitative and objectifying. This question had to be reassessed later on because women vacated the frame and became invisible in a sense. When I made *Measures of Distance*, the video with my mother, I was criticised by some feminists for using the naked female body. I was accused of being exploitative and fragmenting the body as they do in pornography. I felt this was a very narrow minded and literal interpretation of feminist theory. I saw my work as the celebration of the beauty of the opulent body of the aging woman who resembles the Venus of Willendorf – not exactly the standard we see in the media. And if you take the work as a whole, it builds up a wonderful, complete image of that woman's personality, needs, emotions, longings, beliefs and puts her very much in a social context.

(M. Archer *et al.*, *Mona Hatoum*, 1997, p. 41)

About five minutes into the sixteen minute video, the voice-over tells us of the father's shock, during one of the daughter's rare visits home to Beirut, at finding his wife and daughter, together and naked, in the shower. His distress was further compounded by the fact that his daughter was taking photographs of her mother. This exchange of looks, recorded in the images on which the video piece is based, and commented upon in the exchange of letters that recall that moment of mother/daughter intimacy, captures a radical shift, a reorientation between the looker and the looked-at whose hierarchical and gendering asymmetry of knowing subject/ known object have been encoded into Western art as both the elevated artistic nude and the base pornographic 'nudie pic'.

Whose body and whose sexuality have images of the female body traditionally encoded? What rules and conventions govern the visual representation of the sexual body that, during the nineteenth century, came increasingly and exclusively (in terms of both art and the popular media) to be primarily the *female* nude? Who is looking? Who is and is not allowed to look and thus to know the body of the sexualised other or, as importantly in the case of women, the sexualised self? The answer according to those who first posed these questions in the field of art and film histories was crude but not inaccurate: men look at images of 'woman' in which 'woman' is rendered an object of this desiring and voyeuristic gaze, while being fashioned aesthetically as the exhibitionist, connoting 'to be looked-at-ness'. I want to explore what happens to relations of sexuality and visuality when the other, the object become subject, looks at herself, or at the body of the Mother, a fantasmatic imago that quite differently and complexly underpins both the fantasies and the violence of men's representations of women's bodies in both art and pornography? From this third position of feminist interrogation, art and pornography may no longer appear so structurally different, forming only apparently separate ends of a deeply engrained psycho-sexual continuum.

In terms of the scene reflected upon by Mona Hatoum's video, the mature woman's body, naked in the shower, seemed only to become a sexual object as the father encountered the shower scene. He claimed this naked body as *his*, one that should not be seen by any eyes but his and thus should not be *enjoyed* – the verb in both French and English having sexual meanings and connotations of property – by anyone save him. The 'crime' of the exposure of this body lay not in *its* pleasure, in *its* sexuality. The affront arose from the sexual possessiveness of the viewing eye, from the fact that, in fantasy, the eye is an eroticised organ and sight a pathway of not merely desire but mastery that is further framed by a deeply embedded politics of veiling and exposure.

The video, *Measures of Distance*, is exemplary of a current aesthetic and feminist project to challenge the double legacy of what the feminist theorist Jacqueline Rose named 'sexuality in the field of vision' by attempting to represent the female body as the site of her own proceedings, the sign of inscriptions on a cultural text 'in, of, and from the feminine'. This phrase suggests that making a difference to established conventions and meanings around the representation of the female body is not simply a matter of producing alternative 'expressions' that arise from a given gender identity consistent with the anatomical characteristics of the female body: women artists make women's art by expressing a given female bodiliness. From a psychoanalytical point of view, the relation between the physical body and the sexual body is highly mediated by fantasy that cares not at all about perceptual or anatomical reality. What we are does not stem from the given facts of the bodies into which we are born; we inhabit physical potentialities through an erotic zoning that is a fantasmatic composite of libidinally invested boundaries. Much of the process by which the body's potentiality for pleasure and horror is encrypted into a psychic code becomes, by force of a Symbolic law, unconscious, unknown to us while yet affecting us, as we enter into language. The subject, be it the artist who makes or the viewer who reads that art work, *discovers*, in an art work that evokes the bodily by luring vision, something unforeseen about subjectivity and its complex relations to desire and pleasure. Art does not reflect what the subject knows about itself, its sexuality or its body. As an oblique mirror of the subject's formation, artistic representation refracts the hidden memories of archaic experiences and pre-Oedipal fantasies, confirming or shifting our sexed subjectivity at the point where we encounter and experience the image as fantasy.

I have started with a video work that implicitly contrasts one set of cultural conventions surrounding the visual sexualisation of the feminine body through exposing it to sight with the 'inscription' on the video screen of charged and affecting relations between a mother and her artist daughter encountering their feminine sexualities 'in the buff', as it were. They begin to talk and make images that support and refract their conversation about their shared but different female bodies, pleasures, anxieties. Their words concern their own, generationally diverse and culturally located experience of sexuality, motherhood and non-sexual yet almost erotic woman-to-woman intimacy. In traditional Western painting or its modern and popular porno-graphic sister arts, such intimacy would typically be refracted through a voyeuristic Orientalist fantasy of the enclosed feminine space of the harem, where women sensuously prepare themselves and await the summons of the master-lord. Pornography may incessantly show the image of woman. What it advertises, however, is the phallus that, absent from the image, is, none the less, conjured up as always ready to provide real satisfaction that the image/woman structurally lacks. In radical dissidence from that regime of viewing relations, this video allows, to speaking, self-reflecting female bodies, a guarded but luxuriant visibility that, nevertheless, screens it with writing while the video format frames the shifting and shadowed image of the mature body with female voices in the sonorous space of the mother–daughter relation, however strained and pained by the historical predicament of the exiled Palestinian people. It offers, therefore, as radical a gesture with regard to the historical staging of feminine sexuality, generation and cultural iden-tity as any of the major history paintings using the nude did at the trau-matic beginning of modernism, such as Manet's *Olympia* (1863-5, Paris, Musée d'Orsay).

### Where to start, now?

Once a lively and agonistic debate, and still a fiercely fought legal and political battle in the USA, the boundary between artistic representation using the naked body, renamed as nude when clothed by art, and pornography's explicit showing of sexual activities for the purpose of sexual arousal and erotic stimulation is mostly fought out now in the press around art and homo-sexuality (the great scandal of Robert Mapplethorpe's homoerotic photo-graphs of the masculine body comes to mind) and in the courts around the representation of children. Consider the recent cases where photographs by parents taken of young children in their baths have been seized, leading to criminal prosecution of the parents for 'child pornography'. Any looking is now so infested with sexual potential and thus danger that we seem to have totally lost faith in the defence of art, let alone that of motherhood.

My brief was to address the boundaries between art and pornography with regard to *representation*, the invocation of a spectator at the sight of the human body to sexuality, real, potential or sublimated. This boundary lies between an elevated artistic frame in which 'woman' is represented as passively and almost unknowingly participating in an erotic scenario – Venus rising from the sea was a classic convention for showing a beautiful naked female form to which clung the innocence of newborn purity – and scenes in which the represented woman is punished, humiliated or degraded because she is shown both actively participating in sexuality and revealing her fundamental lack in doing so. Critical to this distinction is not the fact of naked exposure, aestheticised or brutally crude. Rather the difference lies in the erasure of any sign of female sexuality from the fantasmatic body that we call the artistic nude (no genitals, no body hair, sometimes no consciousness) and the frank fascination with the genitalised and sexually complicit body in pornography. The absence of signs on the idealised aesthetic body signifies the absence of self-knowledge on the part of the representation of the 'castrated' woman, whereas the explicit depiction of anatomical difference signifies a degradation of 'woman' by sexual knowledge that only further justifies her humiliation. Beneath this binary lurks, of course, the theology of Christianity's deeply divided concept of 'woman' as fallen Eve and sacred Virgin.

My purpose is to set aside this tired question of the nude and the licit versus illicit, elevating versus obscene, representations of sexuality and to explore instead the difficult process of women's reclamation of the representation of the body for feminine subjectivity and sexuality in a space beyond hetero-patriarchal sexuality as it has until now defined the terms of the debate: straight men looking at various fantasies of their psycho-sexual selves called 'woman'.

### Art, pornography, erotica: boundaries?

Perhaps I would have fulfilled my readers' expectations more had I started with some lurid feminist condemnation of the grossest forms of pornographic imagery such as the image from *Hustler* magazine of a woman, spreadeagled and tied with rope on the bonnet and bumpers of a Jeep containing two men in hunting gear, with rifles, with which anti-pornography theorist Andrea Dworkin begins her study of *Pornography: Men Possessing Women*. One of the most outspoken critics of pornography as a part of a system of male power over women, Andrea Dworkin dismisses the distinction upon which I have been invited to dwell, between pornography and what she calls 'erotica'.

> In the male lexicon, which is the vocabulary of power, erotica is simply high-class pornography, better produced, better conceived, better executed, better packaged, better designed for a higher class of consumer. The pornography industry, larger than the record and film industries combined, sells pornography, 'the graphic depiction of whores'. In this male system, erotica is a subcategory of pornography.

(*Ibid.*, 1981, p. 10)

Framed entirely within a heterosexual matrix in which all sexual dimensions are infected by power relations that pornographic imagery plays to and fantasmatically reconfirms, this view of pornography encompasses all sites of sexuality in the field of vision. Is it, however, necessarily true that any representation of the body with an indirect erotic aspect is linked to this deep structuring of sex by masculine heterosexual power? Has not art been fundamentally defined (at least by Kantian aesthetics) as a disinterested contemplation of an ideal beauty that abstracts and sublimates the real body in the interests of an aesthetic experience to tame and and transform any residual sensuality in the name of beauty? Can feminists reject this philosophical premise without abandoning any critical role for art? On the other hand, is not sexuality itself a far more protean and fantasmatic domain in which the relation between fantasy and reality is intrinsically and creatively unstable? Can artists who are women redefine the relations between representation and sexuality in the name of something other than political or traditional morality that must necessarily condemn any use of, exposure of, or address to sexuality in the field of vision?

### The nude: a category art invented

The traditional defence of art against charges of pornography goes something like this. Some images of naked female bodies are defensible because they offer themselves aesthetically reformed to disinterested contemplation, because they are formally reconceived and self-consciously artificed: in a word, they are remade by/as art. Other kinds of images of bodies are indefensible because they are crude depictions of actual bodies and acts and directly aim to arouse sexual feelings. They may even incite sexually related crime.

The definitive art historical text on the genre of the nude as an ideal art form in Western painting appeared in 1956. Published by Kenneth Clark, then Director of the National Gallery, *The Nude: A Study of Ideal Art* offered both a Kantian view of art and a defence of art based on the premise crudely outlined above. Never directly addressing pornography, Clark opened his book with a distinction between the naked and the nude. The nude in art is precisely the opposite of the defenceless and often pitiable human sexual body in its all too physical reality. The nude takes the variable and often very unideal human body to be re-*form*-ed by artistic selection and refashioning as a formal object of contemplation and elevation. Thus the nude relieves those who contemplate the body in art of all reminders of a material, mortal and imperfect physicality, of mortality and hence of sexuality. The painted or sculpted body becomes a kind of fetish, a means of disavowing (a Freudian term meaning 'I know but I don't [want to] know') our precarious and vulnerable humanness. In the form of the female nude, viewing by heterosexual men of the unclothed body of a woman in artistic representation also manages both to satisfy a distantly sexual pleasure and disclaim its overt sexuality. This paradox caused consternation amongst those Christian viewers not versed in Kantian aesthetics, for whom any nakedness was sexual, arousing and thus implicitly erotic or overtly pornographic. Thus the term 'the nude' was coined by eighteenth century art lovers in their attempt to persuade the 'artless islanders' (the British) that 'in countries where painting and sculpture were practised and valued as they should be [Italy and Greece], the naked human body was the central subject of art' (Clark, *ibid.*, p. 1). Not only the central *subject* of that tradition that from the eighteenth century served as the canonised myth of origin for western Europe's assumed racial and cultural superiority, the nude was also, according to Clark, a *form* of art. Just as opera was an art form invented in seventeenth century Italy, so the nude was a form of art invented by the Greeks of the fifth century BCE. Only certain eras and countries, however, could grasp the twin structure of subject and form and produce the real thing: the nude. Some unfortunates continued to create merely untransformed images of the naked body; thus Clark dismissed the northern European tradition of a more realistic theology of the mortal and thus naked body. Greek culture and its apparent revival within fifteenth to sixteenth century Christian societies in Europe under the dominant theology of the Incarnation alone managed to maintain the idea of the body having a meaning beyond itself: being a sign (carrying symbolic meaning) rather than an index.

Furthermore, Clark classified a range of tropes for the symbolic art body thus framed as an ideal form, the very proof of art's alterity and transcendence of the material world of both sex and death. Clark categorised several types of the nude. This is where this gets confusing. For in its history the nude has been far more substantively about a heroic concept of masculinity than about the erotic representation of the female body. There is an important relation between these divergent images of the male and female bodies in art.

Thus Clark started with a male nude, which is the body of masculine identification and narcissistic idealisation, namely the Apollonian body, balanced, prosperous, confident. Here the mind is master over a perfected and creative body. There is also the masculine body expressive of Energy, a less balanced but hyperphallic antithesis of harmonious Apollo typified by Hercules. The body of Pathos for which the Crucifixion functioned as the crucial illustration emerged slowly in Christian society while it also found classical prototypes in the famous *Laocoön* (second century BCE, Rome, Vatican Museum). Finally Clark considered the body of Ecstasy – the Dionysian body that coupled religious possession, drunken physicality and even madness. Within these categories outlined in Clark's book, an interesting statistic emerged when the illustrations were counted and compared. This typology had a gendered dimension. The chapter on Pathos offered thirty-six images of male bodies and only four female; that on Energy had a the ratio of thirty-six to five but, in the section on Ecstasy, the proportion was inverse: nineteen female and only seven male. The mindless body of aroused and disordering sensuality or possession was selectively female in contrast to the pinnacle of the nude as the perfection of a mind/body harmony that was Apollo – not a male *sexual* body but the body as the very sign of self-possession and harmonious subordination of the body to the mind's mastery and to artistic self-reflection. Apparently symmetrical with the figure of Apollo, Clark discussed the origin of the *female* nude in the figure of Venus. It must be stressed, however, that in terms of the development of Greek art and culture, the female nude was not the complement to Apollo and was only a very belated phenomenon. Not invented at the same time as the other male categories of nudity, the female body's artistic unclothing was slow and scrappy until the middle of the fourth century BCE, when the Knidian Venus by Praxiteles appears with its monumental but decisive representation of the woman's nudity as something slyly glimpsed by a prying spectator. Venus is shown as she prepares to enter a bath, clutching her discarded garment, and thus signalling that we, the spectators, are witnessing a moment of surreptitious discovery tinged with panic. Yet Venus is from then on shown gesturing towards her sex with a hand that both attempts to cover and yet points, that both tries to protect and yet underlies that this nudity is about a *sexual* vulnerability to an invisible spectator. In an extraordinary double bind, by this gesture her body becomes only and totally the displaced sign of a sexuality that is, none the less, erased. As the art historian Nanette Salomon has shown in an article entitled 'The Venus Pudica . . .', the gesture and the pose has passed into the Western imaginary to be reiterated down the ages. Thus Clark's supposedly clear division between the body as an artistic signifier for human ideality and the body as mere vegetable mortality is given the lie.

In the chapter on the figure of Venus as prototype for the eroticised female nude, Clark is forced to create two sub-categories: Venus I is Celestial Venus, her form contained by the total erasure of any actual physical signs of woman's sexual anatomy. Enclosed like a bud within a bounding outline, the Celestial Venus culminates in Giorgione's *Sleeping Venus* (ca 1510, Staatliche Kunstsammlungen, Dresden), supine and asleep, the body laid nakedly in a verdant landscape almost becomes itself a landscape (Figure 2). For Clark the opposite is Venus Naturalis (Venus II). Here the bounding outline is breached by a bulging, or ageing, or otherwise imperfect body that escapes its artistic framing and it seems to threaten Clark with a collapse into a vegetal nature that can only recall mortality. A dreadful example would be Rodin's *She Who Was Once the Beautiful Helmetmaker's Wife* (Figure 3; 1880–85, Musée Rodin) or more typically the paintings of known lovers by artists such as Rembrandt and Rubens. Clark's well-dressed art historical cateogorisation could, however, be otherwise interpreted as his masculine anxiety attempting to deflect any return of the maternal body, the body of

103

FIGURE 2. *Sleeping Venus* by Giorgione, ca. 1510.

an opulent, mature sexuality perhaps, a body, which, rather than suggesting the woman's lack covered and protected by the Pudica's hand, is inclined to delight in its own excess: the *Willendorf Venus* (ca 2100 BCE, Vienna Natural History Museum), mentioned by Mona Hatoum, perhaps.

Feminist art historians have had a field day with Clark's unconscious enunciation of a gendered (and classed and raced) position. He is haunted not by the spectre of the pornographic but by that of femininities other to that contained by the subliminally sadistic Greek modelling of the female nude based on the Pudica. This other is obliquely spoken in his text by the term 'the naked' that signifies a kind of unclothedness that is not in reality the simple factitious body, but rather the body unclothed by and escaping the regulating frame of aesthetic re-presentation. Clark calls the nude only the body art has ideally reformed for *his* viewing ease. The feminist critique defines the nude as that which, at the sight of feminine difference, the masculine psyche must refashion to manage the contradiction between a sexual image and the anxiety the otherness of the feminine body incites. Thus, for Clark, artistic representation allows the masculine spectator/connoisseur an untrammelled identification with an ego-ideal that oscillates between rational self-possession (Apollo) and jouissant but still self-affirming abandon (Dionysus). On the other hand, however, idealising representation of the tamed artifice called the female nude relieves the masculine subject of the

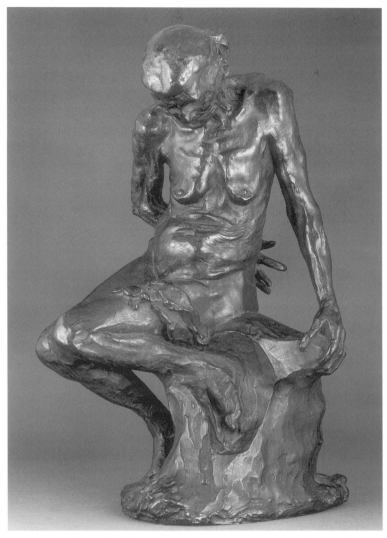

FIGURE 3. *Celle qui fut la belle heaulmière* (*She Who Was Once the Beautiful Helmet Maker's Wife*), by Auguste Rodin, 1880–85.

terror of contemplating sexual difference except through a highly regulated aestheticised script that remodels the female body so that its visual form, bearing almost no relation to the specificity or sexuality of women, now accommodates the defensive masculine fantasies of its makers and viewers.

Is modern pornography – as opposed to traditional illicit erotica – any different? Here is a definition of pornography offered by the Williams Report on Obscenity and Film Censorship (1979):

> a pornographic representation is one that combines two features: it has a certain function or intention, to arouse its audience sexually, and also a certain content, explicit representations of sexual materials (organs, postures, activity, etc.). A work has to have both this function and this content to be a piece of pornography.

(*Ibid.*, p. 103)

Thus pornography differs from art because it aims to elicit the very sexual response ideal art claims to sublimate, and because it does so by showing the signs of sexual anatomy that ideal Western art so extraordinarily erased. Bernard Williams, author of the Home Office Report on Obscenity and Film Censorship just cited, represents the classic liberal position on the line of demarcation. He reconfigures the problem of the permitted visibility of sexuality by appeal to the regulation offered by the law. The liberal calls upon the law periodically to redefine the relations between public visibility and private use of the visual expression of the sexual drive that is considered to be a fundamental term of individual freedom. The first group of images – art – sublimates the sexual, inviting an acceptable form of public contemplation of the indirect evocation of the erotic; the other, pornography, promotes the degradation or commercial exploitation of an active sexuality and thus must be kept from excessive public view while not being condemned, for what we do sexually, within other limits, is part of the freedom enjoyed by the individual citizen. Thus the distinction in effect operates as one between private and public. Some feminists would contest precisely this concept for its inherent indifference to the sexual hierarchy that makes one sex's body both the public and private property of the other's sex's civil rights – without reciprocity.

In this argument, however, art and pornography are conceived as long-standing variations of human sexuality and its representation, sexuality itself being conceived ahistorically and universalistically. (The challenge to this position is best explained by Michel Foucault in his *History of Sexuality: An Introduction Volume 1*.) This position has been vigorously challenged by those who consider sexuality as deeply conditioned by historical changes and fundamentally a social regulation of human potential in the interests of social aims and even class purposes.

In this light, far from being an ancient and honourable tradition of frankness about sexual drives and their enjoyment, which has always been the underside of a beautified and elevated idealisation of the erotic, pornography has been defined as a modern industrial product of nineteenth century technologies of the visual – photography and film – which, in film theorist Linda Williams' words, incited and fed 'a frenzy of the visible'. It was only the exponential increase in the production of sexually explicit imagery made possible by photography, and then film that necessitated legal regulation and public debate about the dangerous excess of the visibility of sexual representation. The distinction between High Art and crude erotica, often in the past made by the same artists who produced the elevating idealisations of the nude, has thus migrated into a modern debate about what can be legitimately shown in public in contrast to what is *obscene*, and, therefore, what should not be seen, even while it is neither morally nor politically condemned per se. The first Obscene Publications Act was passed in Britain in 1857, a date that links the issue's appearance as a legal and economic one to both the development of photography and the expansion of both the popular press and popular forms of public entertainment. For Walter Kendrick, this date also marks a critical moment in the relations between sexual knowledge and power. Linda Williams comments: 'His [Kendrick's] history of pornography, then, is fundamentally the modern story of how those in power react to texts that seem to embody dangerous knowledge when in the hands of the 'other', ... (*Hardcore*, 1990, p. 13). The other in 1857 was middle class young women susceptible to corruption by romantic novels and explicit pictures. Thereafter, the boundary between art and the obscene, pornography, has been moved or challenged in the law courts according to changing social attitudes, administered by 'those in power' through the mechanism of the law which itself now effectively defines pornography as simply those representations of sex that society proscribes according to its varying categories of obscenity, of what should not be publicly shown.

Contesting the erudite connoisseur Kenneth Clark who, none the less, lacked critical self-awareness, John Berger, in his television series of 1972 *Ways of Seeing*, returned to the nude in art to question the radical division based on either of the above distinctions. He argued that there was a continuity between the nude in art and the pornographic nude in modern photography. He wanted to tell it straight. Both are all about sex, and both follow a logic of masculine power in looking at women. Berger compared the overtly sexual come hitherishness of tame 1970s soft porn with images from the canon of Western painting such as Ingres's *Odalisque* (1814, Musée du Louvre, Paris) that functioned similarly to incite the look of the viewer at the display of the female body. While maintaining this continuity, Berger drew his own distinction between nakedness and nudity. For Berger, nudity in any form of a body exposed to another's anonymous gaze via painting or photography is a kind of false clothing imposed upon the body. Nudity as masquerade in a perpetual peep show can be contrasted with the simple fact of nakedness in which one experiences the body as the private and personal site of one's unique social and historical self. For Berger, then, there appears to be a way of experiencing the body outside of all representation, outside of any predetermining schema and this makes our nakedness the irreducible site of a unique individuality and an authentic sexuality. Thus certain images defined by Clark as vegetally too untrammelled become, for Berger, through those very signs of bulging flesh and real ageing that appalled Clark, the connoisseur, the condition of a loving exchange and mutual acknowledgement between painter and model. Paradoxically, Berger can find within art instances of 'naked' painting where the unclothed body is not nude, and does not conform to being dressed in the codes of artistic exposure of Woman for man. In paintings by Rubens of Hélène Fourment and by Rembrandt of his wives, the picturing of a woman's unidealised nakedness can signify a desiring relation to a singular individual. But nowhere in these books does the question of this form and its problem elicit a question about female subjectivity and bodily inscription.

### Men look: women appear. Or feminists use Freud to good effect

From time to time I have mentioned 'feminists' or ' feminism' as one of the major parties to the discussion. Indeed the whole argument was radically realigned by the re-emergence of a self-conscious critical, political and cultural return to the repressed question of gender, the unfinished business of the last great modernisation: the relations between women and men, women and society. Since the 1970s, feminism has become one of the major theoretical and intellectual projects of contemporary critical thought and cultural practice. It has focused extensively on the politics of representation, and within that field on the question of sexual difference.

In a historic edition of the feminist magazine *Spare Rib* in 1973, the film-maker Laura Mulvey published a pioneering feminist response to an exhibition of the work of the British pop artist Allen Jones at Tooth's Gallery in London. Allen Jones' new sculptures formed a now notorious series entitled *Women as Furniture* in which life-size effigies of female bodies, garbed in an array of tightfitting catsuits and piercingly high stiletto heels, cast in a range of sexually provocative and submissive postures, doubled as hatstands, tables and chairs. Allen Jones appeared to be rehearsing the vocabulary of the sado-masochistic porn magazine within the confines of the art gallery. Rather than maintaining its boundary with the pornographic, art appeared now know-ingly to compound the insult of the degradation and dehumanisation of 'woman'. Far from joining in the widespread denunciation of these works for extending the crude sexual objectification of women's bodies from pornog-raphy to Bond Street, however, Laura Mulvey hailed Allen Jones' work for the way in which it *exposed*, through its calculated mimicry, the underlying *psychic* forces that shape the language of phallocentric representation, most visible in pornography yet widespread in the whole field of patriarchal visual representations of 'woman'.

> It is Allen Jones' mastery of the language of basic fetishism that makes his work so rich and compelling. His use of popular media is important not because he echoes them stylistically [as is the character of Pop Art] but because he gets to the heart of the way in which the female image has been requisitioned, to be recreated in the image of man.

(L. Mulvey [1973] in *Visual and Other Pleasures*, 1989, p. 7)

This was a revolutionary revision. What we look at in so-called 'images of women' bears precious little relation to their supposed referent: female people. Rather we are seeing an image that creates 'figures in a masquerade, which express a strange male underworld of fear and desire' (*ibid.*, p. 8). Laura Mulvey's case was built on suggesting that the conventions of Western representation that produced these images and their sources were funda-mentally fetishistic: this meant that images of women could be categorised in the realm of popular pornography of the type Allen Jones both mimed and critically reiterated in three ways: woman with phallic substitute; women minus phallus punished and humiliated often by another woman plus phallus; woman as phallus.

The introduction of psychoanalytical terminology departed from previous incursions by psychoanalysis into art history where the focus was more often than not on psychobiographical interpretations of an individual artist's psyche and *his* related problems. Laura Mulvey signalled a theoretical development central to feminist critiques since then. The object of analysis is the syntax of the representation and the relation between representation and the sexed subjectivity of the presumed spectator.

So what is this regime that renders representation of the image of woman *fetishistic*? The answer is phallocentrism, a term that specifies an organisation of meaning in which the condition of meaning is the set or interrelating binary oppositions: presence/absence; plenitude/loss; one/other. That which organises and is privileged by this binary is the phallus. The phallus is not identical with the actual male sexual organ; the phallus is a symbol, an abstract meaning-bearer that does privilege the masculine sex, who, having a visible appendage, imagines that this possession is identical with that which the phallus signifies: the positive value of presence in contrast to a negative absence (invisible feminine sexual difference). To acquire a sex is to be positioned in one of two relations to the signifying phallus. We can either imagine that one day we shall *have* the phallus (because we have something that can be misrecognised for it, the penis, the phallus appearing always to be in fact the Father's): the masculine position. Or one is forced, because one appears to lack the necessary currency, the penis, to try to *be* the phallus for the masculine other: the feminine position. Either way whatever the difference between the sexes means, it is orchestrated around this asymmetry of having or, through lack, being the phallus of an other. Hence the crisis of masculinity that fears the loss of its privileging relation to the empowering phallus, that fears castration which 'woman', appearing already to have suffered it, perpetually threatens through any display of a 'difference' that is read only as absence [of the phallus]. Fetishism is a defence in which the masculine psyche tries both to recognise woman's lack (hence his *being* in contrast to her lack) while disavowing that knowledge through granting the female body in fantasy a substitute for the missing penis: phallic attributes like shoes, leather, bonds, fur, whips, stilettos etc. These are necessary to reassure the narcissism of the masculine subject that what he 'has' is all and everything that is valuable. Laura Mulvey writes,

> Most people think of fetishism as the private taste of an odd minority nurtured in secret. By revealing the way in which fetishistic images pervade not just the specialised publications but the *whole of the mass media*, Allen Jones throws new light on woman as spectacle. The message of fetishism concerns not woman but the narcissistic wound she represents for man.

(Laura Mulvey, in *Visual and Other Pleasures*, 1989, p. 13)

Turned into objects of display according to a psychic structure in which they are but ciphers of a threatening absence that reminds the masculine subject of the threat of mutilation and loss of self, women are not actually represented there in the wealth of representation artistic, popular or pornographic that offers to us an *image of woman*. Such an image is in fact the indirect representation of masculine anxiety.

> The parade has nothing to do with women, everything to do with man. The true exhibit is always the phallus. Women are simply the scenery onto which men project their narcissistic fantasies [and fears]. The time has come for us to take over the show and exhibit our own fears and desires.

(*Ibid.*, p. 13)

What appears to be an intensely defended boundary between the licit exposure of the idealised body (the nude in art) and the illicit incitement of the sexualised body (the pornographic image) fade before a radically different reconceptualisation of the field of the visual in relation to a psychoanalytically theorised concept of sexuality as fantasy shaped by the logic of masculine castration anxiety whose major defence is fetishisation. The naturalist explanations of sexuality as an inherent force that images may contain or arouse is displaced by a scenario in which pleasures in looking that hover between identificatory looking and looking as mastery – voyeurism or sadism – are related not to some perceptual body but the psyche's reading of sexual difference as the play of presence/absence. The call is not for less sexism or more truth but for other regimes of representation shaped to the measure of as yet unenunciated feminine desire, fantasy, with its own specific fears and anxieties. This will involve both a challenge to prevailing fetishism in the image, and voyeurism in the mode of looking.

111

### Women's images/other desires

To begin to map that project we would have to turn firstly to Linda Nochlin's contribution to the volume *Woman as Sex Object* published in 1972, 'Eroticism and female imagery in late nineteenth century art' and to Lisa Tickner's path-breaking article in *Art History* in 1978, 'The body politic: female sexuality and women artists since 1970'. Note that two terms have entered the arena unashamedly: eroticism and sexuality. These terms indicate the sexual politics of the 1970s in which the crime was not the pornographic versus the elevated but their shared erasure and repression of *female* eroticism or sexuality in both deeply interrelated systems of representation. 'Those who have no country have no language. Women have no imagery available – no accepted public language to hand – within which to express their particular point of view', stated Linda Nochlin in a paper that clearly imagined and condoned the desire to create such a female language of eroticism. Lisa Tickner, however, discerned the necessity for a contrary move: the de-eroticisation and decolonisation of the female body through the challenging of taboos around which elements of that body would become part of the language of art, which exclusions would form the basis of the perpetuation of women's alienation in images of their own absence projected to them by the whole of culture. Since the 1970s there have been many major interventions into this 'country' by artists who are women: Jo Spence, Nancy Spero, Jenny Saville, Laura Aguilar, Nikki de St Phalle, Lynda Benglis to name but a few. Their purposes were diverse but could be connected through the desire to inscribe a different, sexuated desire within the repertoire of representations of the body and the need to negate existing limits within that still phallocentric repertoire. These desires and needs produced two kinds of project: one is to link the body as sign with subjectivity 'in the feminine' and the other to link the body as sign with desire 'in the feminine'. This formulation 'in the feminine' is intentionally tiresome; it aims to problematise any naturalisation of the notion of gender and address the psychosymbolic formation of sexual difference. This involves, above all, discovery of dimensions of what it might mean to be a woman in a culture which signifies – i.e. has recognised signs for – what the 'feminine' might mean beyond phallocentric fears and desires, despite the vast array of 'images of women' that are, in fact, oblique images of masculinity. In a very real sense, we are still 'becoming-women' after aeons of patriarchy.

To underline the enormity of the task of aligning femininity, subjectivity, sexuality and representation, I want to conclude this chapter with a historical digression to a moment in the 1920s, a moment critical to any feminist historical self-understanding as well as critical to recognising that the debate about the body in visual representation that I am addressing is part of the still unfinished business of the modernisation of sexual difference that began not with the suffrage politics of nineteenth century feminists but with the covenant between modernism and feminism in the second decade of the twentieth century. The trajectory of this alliance was stymied by the eruption of fascism and postwar conservativism and it had to wait until the late 1960s for a resumption of its interrupted experiments.

### What a woman cannot dream

Giving a lecture in 1931, Virginia Woolf dramatised the struggle of the woman artist or writer for access to the sense of self required to write or make art. She described a phantom that haunted her and held her back, because she was a 'woman', from saying what she thought. Virginia Woolf named the monitory spectre the Angel in the House, after a persona in a nineteenth century poem by Coventry Patmore celebrating the ideal of bourgeois femininity that Virginia Woolf's own mother, Julia, had lived, and from which she had died, all self-sacrifice and submission. Virginia Woolf fought against this phantom. She tells us frankly that she had to murder her.

> Had I not killed her, she would have killed me. She would have plucked the heart out of my writing. For, as I found, directly I put pen to paper, you cannot review a book without having a mind of your own, without expressing what you think to be the truth about human relations, morality and sex.

(Virginia Woolfe [1931] in *Women and Writing*, 1979, p. 59)

Virginia Woolf continued in her lecture to stress the violence of the struggle for women to gain access to their own intellectual and emotional identity through a creative relation to the sexual body. She evoked an image for her readers of the major impediments encountered by the artist/writer 'in the feminine'. The woman artist/writer is imagined fishing by the riverbank, lying sunk in dreams on the edge of the water, with a rod held out in the deeps – as if fishing in the depths of the unconscious. As she creates, the line slips easily through her fingers. The imagination rushes heedlessly along its flow-ing bed, only to be brutally halted by a sudden smash against something hard and resistant, stunning the flow of excitement by its absolute prohibi-tion. The woman is in distress. 'To speak without figure, she had thought of something, something about the body, about the passions which it was unfit-ting for her as a woman to say. Men would be shocked. She could write no more' (*ibid.*, p. 61). At this point an internalisation of masculine censorship on what a woman can consciously know about sex or her sexual self stuns the flow of creative imagination and polices the deepest moments of exchange between the body's pleasures and the act of representation. Virginia Woolf concludes that, although she did kill off the pernicious voice of bourgeois ideology, the Angel in the House, and find her own voice as a modern writer, she never resolved the telling of the truth about her own experiences as a woman's body: 'I doubt that any woman has solved it yet' (*ibid.*)

### Three modernist case studies

Here lies the problem I am trying to investigate: at the beginnings of European modernism, sexuality was reclaimed from a predominantly moral and social discourse to become a metaphor for a new truth, a new begin-ning, a reconnection with powerful forces of human nature that would signify the escape from the stultifying regularities and disciplines of work and the social etiquettes of the industrial and professional bourgeoisies. Using exper-imentally the aesthetic and materials culled from colonising expropriations around the world invaded by Western imperialism, modernist artists tried to disrupt the boundaries between tired forms of sexual representation – the etiolated classical nude and industrially manufactured pornography – in order to allow sexuality a dynamic role within the making of the new. The patrons as much as the artists indulged in this fantasy world in which the confrontation of the creative masculine artist with his defining truth and freedom was to be imaged in a complex cultural relation to the body of the feminine other that, by being painted under the influence of the styles of non-European cultures, would signify exotic otherness and primal access to this fundamental 'truth of being'. As Michel Foucault would have it, in his *History of Sexuality*, sexuality is not our truth, but, under the bourgeoisie, this facet of the body, its pleasures, health and perversities were represented as the truth of being – and the source of social danger as well as exciting liberation.

At no point, however, was this modernising concept of sexuality imagined as female or feminine. Indeed high bourgeois ideology preached the idea that women had no sex and certainly that respectable women should never know or admit to knowledge of sexual desire – just a higher moral calling to mother-hood and self-sacrifice. The prostitute was feared and idolised precisely as the antithesis of this clean, pure and decorporealised femininity whose space was home and family. By contrast the woman of the streets was the sewer and the ruin. Pornography, as you will remember, was defined at this time through the archaic Greek word for a writing about prostitutes. Thus the oppositions High Art/Low Pornography, Public/ Private that I tracked at the beginning of this chapter now take their place in the extended field of nineteenth century sexual politics.

Yet cutting across this field, and emerging, as Foucault has argued from the dysfunction of the bourgeois family and its regimes of sexuality, psychoanalysis developed as a discourse promising to make sex scientifically speakable and to provide a tour-guide to the dark continent of femininity. Psychoanalysis was initiated by the shift from the medical gaze at the hysterical female body perceived as the unknowing site of its bodily patholo-gies to the attentive listening to the symptomology of hysterical young women's speech. The hysteric, symptom of the malfunctioning of the bourgeois family, cannot find herself within its sexual orders. Disturbing these impossible and unliveable positionalities, the hysteric poses the radically unsettling question of the instability of gender. Through her disturbed rela-tion to conventional sexuality and gender expectations, she appears to ask: am I a man or a woman? Psychoanalysis may be understood as one of the modernisms: psychological modernism or the modernisation of psychology, and further, the modernisation of sexuality. Psychoanalysis allowed sexuality into discourse, relieving it of the blockages that detoured it into neurotic symptoms or hysteria.

Ironically it was Freud's novel writings on sexuality and the unconscious that provided a vocabulary used by the American critics in the 1920s to embrace the drawings and paintings of a contemporary of Virginia Woolf, the modern American painter Georgia O'Keeffe (Figure 4). Using a still sensation-causing psychoanalytical framework, critics looked at her charcoals

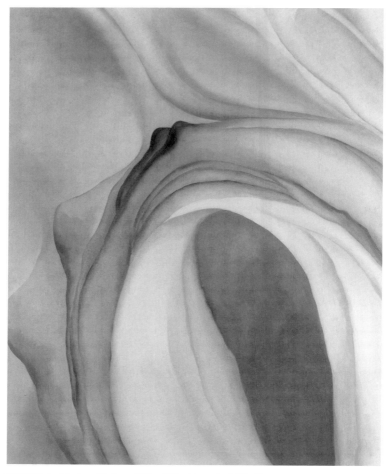

FIGURE 4. *Music – Pink and Blue II* by Georgia O'Keeffe, 1919.

and watercolours and found in Georgia O'Keeffe's work unexpurgated Woman as the pure, unconscious efflorescence of an essential female sexuality. Critics in the 1920s, as well as the American feminist writers of the 1970s, claimed and celebrated O'Keeffe's art as the direct and even revolutionary revelation of what they saw as an intrinsic and corporeal femininity. This was her reputation from the moment Alfred Stieglitz proclaimed on seeing her abstract drawings in 1917: 'Finally, a woman on paper'. In 1924 Paul Rosenfeld wrote that her painting derived from the nature of woman, from an American girl's implicit trust in her senses, from an American girl's utter belief, not in masculinity or unsexedness, but in womanhood. Georgia O'Keeffe gives her perceptions utterly immediate, quivering and warm. She gives the world as it is known to woman ... What men have always wanted to know and women to hide, this girl sets forth. Essence of Womanhood impregnates colour and mass, giving proof of the truthfulness of a life.

(P. Rosenfeld, *Part of New York*, 1924, p. 207)

Earlier in 1922, Rosenfeld had written in *Vanity Fair*:

It is female this art, only as is the person of a woman when dense, quivering, endless life is felt through body, when long tresses exhale the aromatic warmth of unknown primeval submarine forests and the dawn, and the planets glint in the spaces between cheek and brow.

(P. Rosenfeld, *Vanity Fair*, October 1922, p. 112)

In her subtle and intelligent study of three artists, all American, all women, the American art historian Anne Wagner addresses the question of the 'femininity' attributed so approvingly by critics to Georgia O'Keeffe's work in the 1920s. She concludes:

O'Keeffe's art and career also serve to demonstrate that at least earlier in the century, for a woman to represent the body at all is to give herself over to her interpreters; to become hostage to their limited notions of a woman's world as simply and only an extension of her bodily feeling ... I think she continued to see these new pictures as stemming from her long standing effort to find a space both within and outside the body, to conceive of a body both within and outside gender.

(Anne Wagner, *Three Artists* (*Three Women*), 1996, p. 100)

Wagner then cites Maurice Merleau-Ponty as an alternative way to frame an artist who never fully recognised the freight her imagery carried, yet always sought to make her mark lie between its sexuality and its negation: ' il n'y a pas de dépassement de la sexualité, comme il n'y as pas de sexualité fermée sur elle-même. Personne n'est sauvé; et personne n'est perdu tout à fait.' [There is no transcending sexuality any more than there is any sexuality enclosed within itself. No one is saved, and no one is totally lost (*ibid.*, p. 100).]

For the American critics of 1920s, longing for a distinctive American modernism, the work of Georgia O'Keeffe, particularly on paper, offered at the same time what seemed like a direct, spontaneous, almost unconscious revelation of a distinct concept of 'true womanhood' quite different from that which had defined the preceding Victorian or bourgeois culture. For the first time it was possible to conceive of the conjugation of existing ideas of the New Woman, woman changed and modernised, with a degree of sexual experience as the grounds for a new autonomy and self-expression. In contrast to the still dominant bourgeois ideological invention of the sexlessness of Woman, women were being considered as sexual – yet still within a contradictory, phallocentric perspective. As Ellen Kay Trimberger has argued, therefore, in her analysis of the life of O'Keeffe's contemporary and associate Mabel Dodge (1879–1962), a doyenne of Greenwich Village radicalism around 1916, this ambition for sexual self-determination was deeply in conflict with prevailing masculinist ideas of their own liberated sexuality and of notions of woman's differently constituted sexual nature, which was still defined by passivity, receptivity and procreativity. The enthusiasm of some of the advanced critics for Georgia O'Keeffe's work as symptom of this paradoxical notion of woman's unconsciously, almost naturally, sexual nature – hence the floral association – was additionally shaped by then current American appropriations of a simplified Freudian theory of sexuality that was influential after Freud's lectures in the USA in 1909. Mabel Dodge's Salon at 23 Fifth Avenue was a major channel for the dissemination of Freudian ideas in the teens: 'there was no warmer, quiet and more intensely thoughtful conversations at Mabel Dodge's than those on Freud and his implications.'

I want to suggest, however, that after 1917 Georgia O'Keeffe's work did inscribe within the visual field of modernist practice something so visually shocking that only the radical sex talk of the new Freudian discourse could begin to articulate what it seemed to touch upon: a woman as the inscriber of dimensions of difference within sexuality so radically distinct from the existing phallocentric representations of sexuality that it offered the intimations – unknown and unknowable to its author – of a sensualised and corporeal subjectivity that was differently sexuated. So, what was on offer was not 'woman' – an already overfamiliar stereotypical, utterly conventional myth in decadent as much as modernist discourse and representation, where the antinomies of feminine purity and female degradation and vice were regularly recycled. What was glimpsed in her extraordinary images of organic semi-abstraction and then gigantic still life was the possibility that an unsigni-fied dimension of sexual difference could be articulated precisely through the radical, new possibilities of attention to form in avant-garde art. These are not images of the body but images of other elements that evoke traces of barely grasped sensations and semiotic potentialities. Art works that hov-ered between abstraction and figuration through organic analogues and sen-suous colour could evoke an otherness of another sex that is much more than the fiction of 'woman' as the other of the One, i.e. the masculine sex.

But what the critics saw was, none the less, distorted by their own fantasies of the otherness of 'woman'. In 1921 Alfred Stieglitz exhibited forty-five photographs of the younger artist, who had recently become his protegée and lover. The *Portrait of Georgia O'Keeffe* was begun in 1917 when the young teacher visited New York from Texas for a mere three days to see her one-woman show at Stieglitz's 291 gallery. The project lasted until the 1930s when O'Keeffe was photographed with her famous Ford car, the sign of her independence and gradual separation from Alfred Stieglitz, whom she had married in 1924.

In 1921 the grand master of the New York scene presented an exhibition of his photographs under the title *A Woman*. Georgia O'Keeffe is not represented in a studio, with palette and brushes, at work with models and objects, although her body is often juxtaposed to her own drawings and objects. More often, she appears as a body fragmented into elements and parts that in turn attract Stieglitz's analytical and obsessive gaze: face, neck, breasts, torso, sex, thighs, feet, and, always, her refined hands (Figure 5). Paul Rosenfeld wrote:

> [He has] based pictures alone on faces and hands and backs of heads, on feet naked and feet stockinged and shod, on breasts and torsos, thighs and buttocks. He has based them on the navel, the *mons veneris*, the armpits, the bones underneath the skin of the neck and the collar. He has brought the lens close to the epidermis in order to photograph, and shown us the life of the pores, of the hairs along the shin-bone, of the veining of the pulse and the liquid moisture on the upper lip.

(Cited in B. Bulwer Lyons, *O'Keeffe, Stieglitz and the Critics*, 1989, p. 45)

For art historians, and especially those with feminist agendas, this event and its photographic archive present difficult questions. Is this evidence of Georgia O'Keeffe's objectification under the lens of Stieglitz? How could she

FIGURE 5. From *Portrait of Georgia O'Keeffe* by Alfred Stieglitz, 1919.

have submitted to this 'abuse'? Or are we looking at an extraordinary artistic and personal collaboration? Georgia O'Keeffe was a New Woman, politically a feminist, who aspired to the freedom of sexual experience as much as to liberty in her aesthetic practice. In this series, I suggest, that she is seeing herself being seen through the eyes of a heterosexual man who desired her in all the ambivalence in which masculine heterosexualities are forged. The *Portrait* disturbs our common sense because we may feel we are being made into voyeurs who have interrupted a post-coital photo-session. Hardly outside (and even within) the frank depictions of sex we call pornography have we

120

encountered such an evocation of the space and effect of the body as the site and sign of that effect we call sexuality written on and through the body of a woman who must have been conscious of what she was doing when she played her role before the lens of her lover-photographer while he interrogated her body, her skin, her sex for what she dares not imagine.

What would Georgia O'Keeffe have seen in this indirect mirroring? Was it useful as a distancing device, manufacturing the necessary space between the proximity to the body she felt and reproduced in her own work that seemed too revealing and the possibilities that, she glimpsed, might mobilise her experience of her own sexualised and desiring body? This photographic *pas de deux* had, I would like to argue, two authors. The *Portrait of Georgia O'Keeffe* cannot only be the sublimation of the twin faces of desire and tusogyny enacted in traditional imaging of the female body. It is the declaration of intimacy in which the subject, who is not merely a woman-object-of-the-gaze, was an active partner. How radical this was cannot be underestimated. The novelty of the New Woman would lie in that possibility of investing the New Art with foreclosed elements of the feminine subject that an evocation of the body would allow to pass into cultural representation.

By returning to this moment of both sexual and artistic radicalism I have wanted to problematise a simplistic notion of sexualised vision and the body. The case of the Stieglitz photographs of Georgia O'Keeffe and the critical response to O'Keeffe's own work can be complemented by an even more difficult case study of another American modernist whose body became the very instrument of her own mobile modernist dance. Before 1925, the African-American dancer Josephine Baker was marginally famous for comic performances as the girl on the end of the chorus line who got it wrong. Using her sinuous body and labile joints, as well as her expressive features she made her audience laugh. But in one famous dance in *La Revue Nègre* in Paris in 1925, Janet Flanner recounted that:

> She made her entry entirely nude except for a pink flamingo feather
> between her limbs; she was being carried upside down and doing the split
> on the shoulder of a black giant. Midstage, he paused, and with his long
> fingers holding her basket-wise around the waist, swung her in a slow
> cartwheel to the stage floor, where she stood like his magnificent discarded
> burden, in an instance of complete silence. She was an unforgettable female
> ebony statue. A scream of salutation spread through the theatre. Whatever

happened next was unimportant. The two specific elements had been
established and were unforgettable – her magnificent dark body, a new
model that to the French proved for the first time that black was beautiful,
and the acute response of the white masculine public in the capital of
hedonism of all Europe – Paris.

(Cited in Phyllis Rose, *Jazz Cleopatra*, 1989, p. 31)

There is no need to underplay the ocean of European fantasy that swept
over and engulfed this performance in its colonial heterosexual imaginary
when we read André Levinson, the dance critic, writing of Baker as
Baudelaire's Black Venus: 'The plastic sense of a race of sculptors come to
life and the frenzy of African Eros swept over the audience. It was no longer
a grotesque dancing girl that stood before them, but the black Venus that
haunted Baudelaire.' As Henry Louis Gates Jr and Karen Dalton in *Critical
Enquiry* (1998) have written of the shock; 'No one had ever witnessed such
unbridled sexuality on stage. Words like *lubricity, instinct, primitive life force,
savage, exotic, bestiality* and that particularly loaded word *degenerate*, raced
through the capital.' Shock at the display of sexuality – registered through
the racist troping of a modern African-American jazz dancer as an African
sculpture come to life, a primitivising European Pygmalion called into flesh
and bendy bones this African Galatea while not noticing that it was a dance,
a routine, an image, fabricated by a city girl from East St Louis in a complex reclamation of the very terms of her ancestors' enslavement and abduc-
tion from an Africa she had never known. A dangerous game – playing with
fire, we might say. In one of her many autobiographies, Josephine Baker
provides some important glosses on this breakthrough in her career and its
cataclysmic moment of cultural revolution. Paris was the first city in which,
as an African-American, she was served in a café by a white man who called
her *Madame*. She also describes her sessions with the artist Paul Colin com-
missioned to do the poster for the show *La Revue Nègre* described above by
Janet Flanner (Figure 6). In the painter's studio, under the compelling gaze
of the masculine artist whose eroticisation of the scrutiny of the living woman
he asked to unrobe before his hungry eyes has been the object of feminist
obloquy, Josephine Baker experienced a coming into her own sexuality
through the gaze of this other, this white man, this European artist. The
puritanism inflicted on African-American woman in defensive response to
the atrocious abuse of their labouring and sexual bodies by European slave-

FIGURE 6. *Josephine Baker*, plate by Paul Colin from *La Tumulte Noir*, 1927.

owning men was unexpectedly displaced through a scenario that held open the door to Josephine Baker's experience of her own sexual beauty and potential: 'under his (modernist) eyes, for the first time in my life, I felt beautiful'.

This might seem so trite: a black woman finding herself in the sexual glance of a desiring white man. Yet it was under the eyes of an artist, a white man, a Frenchman, that Josephine Baker, like Georgia O'Keeffe regarded by Alfred Stieglitz and his camera, found a look coming from an/Other – a politically problematic place – down which to glimpse another vision of herself, as a sexual being, as author of that charge, that inevitably narcissistic reflection that folds into sexuality. The so-called masculine gaze, institutionalised in art and cinema, that feminists have identified only to decry, could have other uses. It could be instrumentalised as a visual pathway in which another reflection could bring to the surface something that Virginia Woolf, still the child of that Angel in House, found barred and blocked: a 'truth' – or rather an access to something necessary about her own experiences – of her own desiring and pleasuring body. As Josephine Baker describes it: she found someone. She found a 'she' that had always been there although she, Josephine, busy making audiences laugh at and not kill her black femininity, had not been able to see 'her'. 'She' and 'her' signify a resource in black femininity and sexual subjectivity that could fuel Baker's authoring of her own modernist performance art through the reclaimed instrument of her *moving* body in dance.

This moment in the studio of Paul Colin echoes the photo-sessions in New York and in doing so allows another way to elaborate my argument about the distance between the ideological inscription of a nature to 'woman', called her essence, which she must passively embody, and the idea of a masquerade. The masquerade, as Mary Ann Doane has argued, can also be the aesthetic fabrication of a distance between this ideological 'nature' and the subjectivity of the creative New Woman. In order to become not just a political subject, via the suffrage, or political rights, but a Subject, the at least partial author of oneself, through the creation of both a personage and an aesthetic practice that can affirm and enunciate a woman's singularity, women needed a means to articulate a history of the body – the body as space of phantasy, of pleasure, of anguish, the locus of loss, of memory and objectless desire.

Josephine Baker was at that moment a black American, not an African. Yet she wrenched from the negrophiliac Europeans the terms of her own self-invention and self-projection. She reclaimed the black body that had been coloured and colonised by Gauguin and Picasso, rendered primitive and mindless. She animated the body with the musicality of sex of which she was the choreographer and author. Balanced precariously between gross racism and the possibilities of that époque created in colonialism's ambivalence, Josephine Baker's dance at the Théâtre des Champs Elysées was a historic moment for the creation of female sexuality and its aesthetic signification that interrupted all that lay along the boundary that traditionally divided the artistic from the pornographic in the politics of representation and sexuality.

Between the culturally policed and legally defined division between art and pornography, I have tried to weave other stories and trace a critical interruption of the plot I was invited to rescript. On the larger map of historical modernity and its cultural upheavals as well as political tragedies, I have marked a few moments of encounter: between artist daughters and their present/absent maternal imagoes (Mona Hatoum and Virginia Woolf) and between radical modernist women and heterosexual masculine artists who pictured their bodies (Georgia O'Keeffe and Josephine Baker). By allowing us to see how complex is the question of 'sexuality in the field of vision' these image-traces of that encounter await a different reading, a reading that desires difference and allows us to acknowledge how massive is the task, how great the beginnings and how exciting the prospects when women speak, see and represent in their own names and from the site of their own curiosity and desire for their own bodies.

FURTHER READING

Archer, M., Brett, G. and de Zegher, C., *Mona Hatoum*, London: Phaidon Press, 1997.

Berger, J., *Ways of Seeing*, Harmondsworth: Penguin Books, 1972.

Clark, K.,*The Nude: A Study of Ideal Art*, London: Penguin Books, 1956.

Doane, M. A., 'Film and the masquerade: theorising the female spectator', *Screen*, **23** (1982), nos. 3/4, 74–88.

Dworkin, A., *Pornography: Men Possessing Women*, London: Virago, 1981.

Kuhn, A. *Women's Pictures: Feminism and the Cinema*, London: Routledge, 1982.

Kuhn, A., *The Power of the Image: Essays on Representation and Sexuality*, London: Routledge, 1985.

Mulvey, L., 'Visual pleasure and narrative cinema,' and 'Fears, fantasies and the male unconscious,' in *Visual and Other Pleasures*, pp. 14–26 and 6–13, London: MacMillan, 1989.

Nead, L., *The Female Nude: Art, Obscenity and Sexuality*, London: Routledge, 1992. [This book offers an important critique of Clark, *ibid.*, esp. pp. 17–22.]

Nochlin, L., 'Eroticism and female imagery in nineteenth century art' in *Woman as Sex Object,* ed. T. B. Hess and L. Nochlin, pp. 8–15. New York: Newsweek: Art News Annual, 1972.

Griselda Pollock, *Vision and Difference,* London: Routledge, 1988.

Griselda Pollock, 'Inscriptions in the feminine' in *Inside the Visible: An Ellipitical Traverse of Twentieth Century Art, in, of and from the Feminine*, ed. C. de Zegher, pp. 67–87, Boston: MIT Press.

Rose, J., 'Sexuality in the field of difference', in *Sexuality in the Field of Vision*, pp. 224–233. London: Verso Books, 1986.

Salomon, N., 'The Venus Pudica: uncovering art history's 'hidden agendas' and pernicious pedigrees', in *Generations and Geographies in the Visual Arts: Feminist Readings*, ed. G. Pollock, pp. 69–87, London: Routledge, 1996.

Tickner, L., 'The body politic: female sexuality and women artists since 1970', *Art History*, **1** (1978), no. 2, 236–249; reprinted in *Framing Feminism*, new edn, ed. R. Parker and G. Pollock, pp. 263–276 and 336–339, London: Pandora, 1995.

Williams, L., *Hardcore: Power, Pleasure and the Frenzy of the Visible*, London: Pandora Books, 1990.

Woolfe, V., 'Professions for women' (1931), in *Women and Writing* ed. M. Barrett, London: Women's Press, 1979.

## 7 Body, Cyborgs and the Politics of Incarnation

BRUNO LATOUR

> Our body itself is the palmary instance of the ambiguous.

William James, *Essays on Radical Empiricism*, 1912

### Introduction

You may remember a science fiction movie – *Fantastic Voyage* – in which a group of human scientists become so tiny that they can penetrate the organs of a coma patient, traveling through the blood vessels as if rafting through the Grand Canyon, watching with surprise the gigantic machinery of the pumping heart, which, because of the comparative size of the visitors, resembles a huge industrial plant as big and complicated as a nuclear reactor. Indeed, for these minuscule travelers, the body becomes a surprisingly technical landscape, as if they were flying over a densely populated industrial region. Thus, at a certain scale, the distinction between machinery and bodies, individuals and populations, disappears and so does the apparent unity of our body, which is only the superficial impression left by the routine of life. Those who are fortunate enough to enjoy the benefits of good health have one type of body, all the rest have ... what shall we call that proliferation of machinery, populations and spare parts watched by our diminutive futuristic voyagers? Let us call it cyborg, since I was assigned the difficult task of writing about a subject in which, as you will see, I am in no way a specialist.

'Cyborg' is a hybrid term, half cybernetic, half organistic, that was made to define the prosthetic character of our post-human, post-modern, end-of-the-century existence: half a physiological body, half a high-tech robotic one. For me, the epitome of these cyborgs is the dream of Hans P. Moravec, a professor of robotics, who explained to me one night (when we were both celebrating the birthday of HAL, Arthur C. Clarke's *2001* computer) that we will soon be able to download ourselves into a better, drier, 'silicon platform', so as to escape the computational limits of our present 'wetware platform' (his name for the body), which he found to have been inadequately designed to compete with advanced web-crawlers! The idea that humans are computational machines soon to be able to download themselves from their present platform through a modem to compete with virtual reality avatars is the stuff of cyborg fantasy, the pipe dream of technophiles. But to this technophiliac, science fiction, masculinist cyborg has been added, in particular by my friend Donna Haraway, a feminist, polemical, political meaning that is almost exactly the opposite: a designation of the uncertainty about body limits, body fates, body components. The feminist movement has payed a very high price to be able to see, in bodily attachments, not a matter of fate and common sense, but one of emancipation and choice. So, we have one type of cyborg that embodies rationalistic dreams of detachment out of our flesh, and another type that directs our attention to our various attachments, to what could be called the politics of incarnation.

In this chapter I want to look at the second meaning of cyborg, that used to define the front line, or, more exactly, the highly contrasted pattern of many front lines, like the board of an already well-advanced game of *Go*. The somewhat monstrous word cyborg thus denotes a war zone: that of body politics, meaning both the politics of the body, what Michel Foucault called the resistance to biopower, and the body politic, meaning the health of this artificial assemblage that we call a society. The reason I welcomed the assignment given to me by the scholars of Darwin is that I believe there is nothing more urgent than mapping the war zone of what will be one of the crucial issues of practical politics in the next century. A question incomprehensible, even to our parents, is now on the agenda: What is it to bring democracy to the question of having a body? Even more provocative: what is it to have a democratic body? (A vision flashes through my mind: one political cyborg among many – the surgically transformed body of Boris Yeltsin, hooked up to dozens of life-saving instruments, this plumbery of artificial arteries and veins, presiding over a devastated country. Other nightmarish and not so futuristic visions: a patient with the liver of a pig; pellets of cells from farmed human embryos cultivated in cows before being infused into a human brain to fight Parkinson's disease; cloned bits and pieces of organs bubble softly in the basement of a hospital whose cost is sending the social security system down the drain.) So, a new question presents itself: what sort of body do we wish to have? What sort of body is worth having? In the past, this question was raised by religion, education and ethics only about the soul; now it is also raised in the context of the body, by medicine, science, economics, politics and morality.

## The body divided

Let me start with an anecdote: my former colleague in San Diego, Professor Paul Churchland, used to carry in his wallet a picture of his wife, herself a famous neurophilosopher. Nothing unusual in this, you will say. No, except that this picture was an image produced by computed tomography, a CT scan of his wife's inner brain, in full colour! If I had had to go with my wife to the doctor, would I be carrying those cyborgian photographs of color-enhanced neurons? Normally, it seems to me, I would carry my partner's photograph. But Paul was adamant that in a few years, according to him, everyone will become used to seeing neurons, speaking in neuronal terms, and will abandon the comparatively vague language of visage, face, eyes, lips, mouths and subjective portraiture.

We could of course imagine the screams of horror uttered by people engaged in the battle against the naturalization, 'biologization', mechanization of the body. Think of what Emmanuel Levinas, that great commentator on visages, might have said of Churchland's high-tech Veronica, his true image of his wife? Levinas would surely have branded it as reductionist. But this is not the direction I want to take. After all, historians of photography will tell you that to carry a portrait of your sweetheart is a rather new fashion that is itself dependent on the development of a large industry and on the transformation of customs, style, affection and family relationships that is every bit as artificial, historical and mediated as the one Paul Churchland is advocating. So, there is nothing especially subjective, obvious or immediate in 'normal' photography using silver nitrate; nothing especially objective, far-fetched, artificial or reductionist in plugging the language of affection straight into that of acetylcholine receptors, hypothalamus, and ganglia. On the contrary, this anecdote allows me to set aside one type of body battle, that I think is superficial and has been going on much too long. This obsolete fight seems to me to put in opposition the common sense subjective appraisal of human face-to-face encounters and the objective scientific outlook that considers only the physiological basis of neuronal action. The distinction between what could be called a phenomenological body and a physiological one is useful to re-run the tired old diatribe, the gap between the two cultures, but it makes no sense any more because, as I will show, it has inherited a political agenda that is totally outdated in our cyborgian times.

The great advantage of writing for a Darwin Series is that any readers are likely to agree with me about the non-sense of counter-poising the language of face recognition and that of neuronal recognition, since Charles Darwin himself, and many a good sociobiologist after him, have shown how extraordinarily adaptative is the facial expression of emotions in human and animals, especially primates. So very quickly the dividing lines between phenomenology and physiology break down. If you talk to a historian of photography, the photographic portrait soon stops being the expression of an obvious 'subjective' attachment and becomes, on the contrary, a highly social, industrial and cultural achievement. If you talk to an ethologist, the language of face recognition will become exactly as 'objective' as that of its neurological counterpart, and will take you into a history of evolution every bit as complex, material and decisive as that of the electric storms firing into the brain. And what about the CT scan itself? I can imagine many colleagues of Paul Churchland who would carry in their wallets other images that could render his version of his wife's brain exactly as particular and romanticized as the old photographs of the former narrative of affection. What pictures of their 'significant others', are carried in the wallets of Francisco Varela, Gerald Edelman or Steven Pinker, to take a few of Churchland's contradictors or colleagues? I do not think they would agree enough among themselves to offer to the physiological body an indisputable bedrock on which to project the lived-in feelings of the romantic body. So we do not have only one image for phenomenological body and another for the physiological body, but many images of our beloveds' faces and we might want to reclaim all of them; even if we wish to rank them in some sort of order, there is no use in throwing them into two piles – one labeled indisputable science and the other feelings.

You see, I hope, where I am heading? Do not count on me to rehearse again the opposition between a subjective or a socially constructed definition of the body, and an objective definition, and even less to place the warm but vague spiritual lived-in body in opposition to the cold but precise material one. No, the opposition I want to draw, using the tale of Paul's CT scan, is entirely different; it is just as sharp, but cuts through the old party lines. Remember that Paul Churchland did not simply add the CT scan to normal photographic images: he claimed, in addition, that the language of neuronal states will replace, from now on, the romantic discourse of face recognition. He is, in his own terms, as is his wife, an eliminativist. Now here is the problem I want to tackle: Paul does not simply say that CT scans have added to the world a new instrument with which we can become sensitive to differences in brain states that previously we could not have dreamed of watching because we had access only to facial appearances. No; he also says that the language of brain activity will definitively and irreversibly replace the language of common sense affection, or, more exactly, will displace this language, into the realm of subjective feelings.

What I want to do is to disengage the new version of the body offered by Churchland from the eliminativism that is, so to speak, added on to it – for reasons entirely unrelated to good scientific practice. To do so, I want to resort to a distinction, traditional in philosophy: that between the primary and the secondary qualities. This old scholastic divide was used by Galileo, Descartes and Locke to justify the doctrine of eliminativism: for instance, colors and smells would once have been considered secondary qualities (this is no longer the case, which is interesting in itself, since it means that there is a history of the partition of qualities); extension and geometrical shapes, on the other hand, were defined as the primary qualities, the stuff from which nature is made. When Paul uses the CT scan in an eliminativist fashion, he is in effect defining neuronal states as a primary quality, the stuff from which action and thought are made; he is dismissing the rest into the realm of secondary qualities, i.e. the realm of feelings that refer not to what exists objectively but to our subjective way of perceiving, erroneously in his view, the state of affairs.

This is what A. N. Whitehead, in my opinion the greatest British metaphysician, called 'the bifurcation of nature'. Once someone indulges in the distinction between primary and secondary qualities, it is very difficult to stop the warring factions: primary qualities define what the world is like, although the invisible is rendered visible by scientists, whose work remains itself invisible; secondary qualities, on the other hand, define what the world is felt like, what is visible, what is meaningful, but what, unfortunately, is also meaningless because it lacks the essential ingredient of meaning, i.e. being connected to what really exists. Such is the dramatic result of this war that I propose is now obsolete: invisible visible reality on one side and meaningful meaningless representation on the other. The expression 'the making of subjectivity' is at the basis of what I call the true war of science that has divided every fact, every theory, every hospital, every part of nature and, literally, every body in some parts of the Western world. Subjectivity is not a given but part of the boundary dividing the primary and secondary qualities: what the world is like and how we perceive it.

The question I want to tackle here has thus now become: what body politics would not use the primary/secondary quality bifurcation as its main dividing line? To approach the question in this way, you should accept that the very distinction between primary and secondary qualities is not simply a given, but is, on the contrary, a highly contentious compromise: I am not politicising the body, I am placing in opposition one aspect of body politics that I find 'unhealthy' with another that I find more healthy, more democratic, and which will provide, in my view, a better ground rule for adjudicating claims about biopower.

To clarify this point, I want to take another example, more dramatic and more enlightening than my story of the CT scan: I want to look at the AFM, the French Association for the Fight against Muscular Dystrophy, an association with which I have strong links, since my colleagues Michel Callon and Vololona Rabehisora have studied it for several years now.

The reason why this example is so striking is because it is a patient-run organization that has managed, through public charity, to raise enough money to engage in a vast program of molecular biology, with the aim not only of finding the genes responsible for the rare so-called orphan diseases but also of finding gene therapies for them. Genes, as Marylin Strathern has so beautifully documented, are a highly contentious topic. They are often seen as one means of determining human behavior and as such have become a contemporary synonym for destiny and ineluctable fate. Where genes enter, liberty flees away, and I am sure you are familiar with the endless headlines about the tug of war between nature and nurture, the harsh necessity of gene action against the illusion of emancipation. The AFM example is entirely different. This is a case where patients have fought hard, often in the face of opposition from physicians and many geneticists, to discover the gene responsible for their disease. Far from opposing genetic determinism to achieve emancipation, this organization has fought to impose on molecular biologists a research program that would emancipate the patients, through the knowledge of the genes: they provided geneticists with the family trees of their members, they financed the development of robots to draw the first physical map of the human genome, and they are now financing, in the south of Paris, a whole industry around genetic markers and, they hope, gene therapy. Thus, for AFM, gene action is a synonym for emancipation; the parents still remember the time when doctors, in the name of science, discouraged them from doing anything about their dreadful prognosis: 'Your child is condemned to certain death; there is nothing you can do; there is nothing we can do'. Through the patients' action, doctors have been forced to learn a different science of handicap, and French molecular biologists, very reluctantly at first, have entered into an industrial-scale crash research program they would never have carried out without AFM money and pressure.

Clearly, this second example does not toe the familiar party line, where doctors, physicians and biologists are portrayed as defining the real, albeit cold and piecemeal phenomena of diseases, while patients, immersed in their suffering and subjective world, strive for more holistic, humane and charitable treatment. The real fight created by the AFM against the old medicine of resignation does not pit physiological against phenomenological body, cold and objective against warm and subjective, piecemeal reductionist approaches against holistic visions, but the production of resilience against what could be called the fabrication of potentialities. In this new fight, much more crucial, it seems to me, than the old one, science is not innately always on the side of good; more specifically, each discipline, each scientist, each piece of science has to make a choice: to define which side of such body politics it is taking. In the second part of this chapter, this is the problem I want to map out.

### The body shared

What I propose to do now is to move away from the contest between primary and secondary qualities (or facts and values, or the realm of necessity and that of freedom) towards a more rewarding dispute – one concerning our shared world. I want to ask what that world is, without resorting to the traditional solution: it is only the world of primary qualities. In these cyborgian times, this answer has become unacceptable for scientific as well as political reasons. Why? Because it supposes the problem already solved by 'invisible' scientists who seem to be largely unaccountable. When politics, values, or religions enter the scene, it is too late: the backdrop has already been defined by primary qualities, the shared world has already been designed. Apart from nature, what is left are only subjective, cultural, or social representations that are innately multiple, divisive, irreconcilable, incommensurable, meaningful maybe, but also meaningless, since they lack the ability to focus on a shared world, to access with certainty the world out there as it is. Let us call this traditional differential mononaturalism and multiculturalism: one nature interpreted by many cultures, i.e. one unifying nature – the common world, primary qualities – and many divided and divisive cultures.

Common sense tells us that there is no alternative to this differential, the only one we have, the safest we have found during the course of a terribly troubled history. Common sense however, is not always such a sure guide when times change dramatically. For one thing, it would not have helped M. Barateau, the president of the AFM, since his quest for genetic determination and gene therapy would never have got started if he had let the physicians and many of the geneticists of the time dictate the difference between objective and subjective viewpoints. To further the aims of his association, he could not take science for granted; he had to pursue the efforts of his charitable body from inside scientific controversy, sorting out which geneticists to believe and which to disbelieve, whom to hire and whom to dismiss, whom to follow and whom to ignore, thus forcing the professional body of physicians to divide and realign along new loyalties and missions. In that respect Barateau is not alone, as progress in AIDS research shows equally vividly. More generally, as I have often shown, for many of today's new situations, in genetics, ecology, health care, computer science, security, economics, science is no longer what stabilizes the political, moral or ethical contests of normal life, but may even make them flare anew. Plato, in his *Gorgias*, ridiculed democracy because, he claimed, if you ask children to choose between the bitter-tasting medicine of the doctor and the sweet desserts of the cook, the fickle Demos would of course vote the cook into office and dismiss the doctor, amidst much laughter and jeering. A fine story this may have been in the polemical days of classical Athens, but what should we say today when we are faced with many doctors and many cooks who tell us that the new existential question is 'to eat or not to eat beef'? We would be extremely happy if we could even distinguish science from gastronomy and thus be faced with an easy choice between a doctor and a cook. As we all know too well, this is far from being the case. Instead of putting water and sand on the fire, an appeal to science is now seen as pouring oil on the political flames. However, this is a good point at which to try to modify the traditional differential, to see what might have gone wrong in mononaturalism–multiculturalism body politics.

I am well aware that here I am treading on dangerous ground, walking on quicksand, because it seems at first that there is no alternative to one nature except the solution of cultural relativism. First let me take another example to see where an alternative might lead us. The fiftieth anniversary of Simone de Beauvoir's *Deuxième Sexe* was a good occasion for the French to re-enact the old battle between nature and freedom. In *Le Monde*, inflammatory articles saluted the courage with which Beauvoir broke the sway that nature had on the definition of women's existence, and sociobiologists were ridiculed for their attempts to bring culture and emancipation back into the narrow constraints of nature. (Do not forget that in France 'sociobiology' is a bad word, almost as bad as Darwinism, and that it has become the property of the far right. I should add that, in France, I am the only sociologist who has dared to show long-term interest in, and respect for, sociobiological literature.) Now of course, when one reads those appeals for emancipation and hears the equating of the whole of nature with determinism and what Jean-Paul Sartre called 'pratiquo-inerte', the first thing one wants to do is to throw into the discussion a few hormone levels, a good bundle of genes, evolutionary data on the sexual division of labor, and more than a fair share of comparative animal studies, from baboons to ants. It is a sort of knee-jerk reaction against the absurdity of defining emancipation as what escapes from the sway of a homogenized conception of nature. Sure enough, counter-articles by feminist philosophers flooded the press with arguments for the bringing back of the body, for the importance of maternity, for the superficiality of the denial of sexual differences.

That is what body politics right now means: a large part of the dispute is due to equating the appeal to nature with eliminativism. If you hear a French Sartrian feminist pouring scorn on biology, you are bound to retort with masses of biology just to make sure that the shared world is not instantly thwarted and squeezed into a much too narrow definition of freedom. (Remember that, for the AFM, emancipation was gaining the right to identify genes and they managed to force surgeons, for instance, to learn how to practice tracheotomy, instead of letting their children die a 'natural' suffocating death.) Now, of course, the wind blows the other way: if you listen to Hans Moravec declaring that humans will have to migrate to the World Wide Web in 2010, or maybe 2015 (he was not entirely sure), or if you hear yet another talk on what Evelyn Fox-Keller calls the 'discourse of gene action' as if genes were masterminding biological development, or if you hear the Churchlands expounding their eliminativist doctrine once again, you are bound to retort that things are surely more complicated and that the prejudices of those scientists should be carefully counter-balanced. Do you see what I mean? It is only the outrageous position of your adversaries that will force you to take in turn an outragingly false position, simply to resist being squeezed out of existence. This has nothing to do with a struggle between biology and morality, science and society, facts and values, body and soul, but it is a thoroughly political dispute on how we are to define the world that we share, i.e. the right body politic: who is going to assemble the Demos?

Let me underline this point: I am not rehashing the nature/nurture argument, pointing at a more serene, ecumenical, integrated, holistic vision of human existence, which would include as people often say 'not only' biological factors 'but also' subjective, social, human and cultural ones. There has been, in my view, an overdose of 'not only ... but also' balancing acts. The reconciliatory talks are almost as bad as the polemical ones. Nay, they are twice as bad, since they add all the absurdities of one position to those of the opposition; they pile artefacts on top of artefacts. I might not wish to have my body defined by a Sartrian freedom-fighter, nor by some sort of Dawkinsian selfish-gene fundamentalist, but the last thing I want is to have a holistic body half-Sartre and half-Dawkins! That would be a cyborg of an extraordinary monstrous and nonviable variety. No, what I want is to escape from the limitations of these disputes and to extract yet another type of cyborg, some of those 'hopeful monsters' that we could collectively learn to educate.

Before I proceed further, let me emphasize that the Science Wars that have raged (in the press that is, because for the rest of us it was more a storm in a tea cup) has exactly the same quality as those archaic disputes between necessity and liberty. Since those two sham-wars are connected and since both are dealing with a new link to be forged between science and politics, it might be useful to make a small point about its importance. When you hear scientists define a state of affairs as an incontrovertible 'matter of fact' and speak of the indisputable laws of nature, it is quite reasonable to ask: where is the apparatus of those scientists; which laboratories do they work in; how much money do their matters of fact cost; what theories have shaped their selection of data; who supervised their postgraduate theses; what discipline do they pertain to; how many citations in scientific journals do they get; which school of thought, which clique, is holding them in its narrow confine? Plebian questions? Too mundane? Too down to earth? Too bodily incarnated, to use the title of a fascinating book by Chris Lawrence and Steve Shapin? Maybe yes, but this is also a normal reaction, a healthy reaction, against the voice from nowhere that claimed to make nature speak without any intermediary. Of course, here again it works the other way, and if you hear sociologists of science insist that nature does not speak herself and that scientists do all of the speaking, then you will want to rush into the war zone and challenge those social constructivists to demonstrate whether or not gravity is socially constructed. All of that is healthy, normal, human – but is it interesting? Is it important? Is not the question entirely different and the solution lying somewhere else entirely? Should we not let those people react and counterreact, battling among themselves while exploring a more solitary but more rewarding alternative solution?

One issue is recognizing that it is very difficult to be reductionist and here I want to use sociology of science to throw some light on body politics. Let me take the paper by Daisuke Yamamoto (one of our students has been doing field work in his laboratory) on sexual orientation published in the *Proceedings of the National Academy of Sciences* in 1996. The press releases claimed that here was the discovery of 'a gene for homosexuality'. Is the paper on humans? No, on fruit-flies! We are talking here about homosexual *Drosophila* and of a mutation called *satori*, a Buddhist term designating nirvana in Japanese. Typical reductionism here, one might say: a highly complex historical and cultural way of eliciting sexual orientation brutally replaced by the masterminding of one single mutation. The point I want to make is that there is a great difference between claiming reductionism and achieving reductionism. Please do not misunderstand me: I am not fighting reductionism as if there were a better, higher, more generous, more holistic way of doing science. I am disputing that reductionism is achievable at all, if by this we mean the possibility of one level of reality eliminating the next, or for a new entity discovered in a laboratory to erase, hide, or efface its scientific discoverers. The real cyborg in which I am interested is the one made by scientists and their laboratories and their matters of fact: to be sure Yamamoto's paper claims to be reductionist and to establish a direct connection between a gene and a phenotypic action, but when reading it you find a bewildering series of levels, instruments, and oratory precautions that belie the official argument of the text. In turn, natural history, ethology, anatomy, statistics, cytogenetics, molecular biology, DNA databanks, and neurobiology are each used, layer after layer, to establish a highly complex conduit between one mutation and a possible change in sexual preference. The paper concludes:

> Sex is determined by diverse mechanisms that vary among species. Despite fundamental differences in sex-determination mechanisms, the physiological outcomes of dysfunctioning of these mechanism can be quite similar in different animal species. The present findings raise the intriguing possibility that mutations in a single sex-determination gene result in sexual transformation of a subset of brain cells, thereby altering sexual orientation in the male.

If you read the second sentence, it is a done deal: he who holds one mutation of one *Drosophila* fly holds the key to homosexuality in all species that are basically 'similar'. If you read the third sentence, however, there is an 'intriguing possibility' to connect half a dozen mediators from gene expression to brain sites to sexual preference, each of the steps offering a potentiality. That is, each step can fail, not only because it can be, and was, attacked by other *Drosophila* scientists (whom Robert Kohler has called 'Lords of the Fly'!) but also because it might not generate the kind of necessity warranted by the 'discourse of gene action'. Each scientific discipline is thus divided internally between one pathway along which flows indisputable necessity and another that deals with the same reality but mediates it through fiercely disputed potentialities. Let us imagine the following thought experiment: color in red all the points in the reasoning where scientists claim to reveal indisputable necessity that allows them to do away with levels of realities limited to the role of mere intermediaries; for the same reasoning, the same results, the same data, now color in blue all the points where scientists add potentialities, connecting together those mediators that are unable to eliminate the ones following them. I claim that the question would be no longer be 'Is this hard science or hermeneutic?' or 'Does this pertain to physiology or to phenomenology?' but 'What is the color of this piece of science?'.

My contention is that it is always counterproductive to attack or to defend reductionism, as long as you do not make clear which of these two pathways, the blue or the red, you are attacking or defending: is it the second sentence or the third of the above quotation? Is it the claim to connect gene and homosexuality, or the adding, in one Japanese laboratory, of 'intriguing possibilities'? Is it the elimination from the world of any other cause of sexual preference, or the addition of another species, *Drosophila melanogaster*, to the human debate about sexual orientation? Is it the imposition of primary qualities or the proliferation of entities with which we have to share our world?

Many years ago, when I was studying Roger Guillemin's laboratory in San Diego, I was amazed that my humanities friends on the other side of campus could call his neuroscience reductionist when I was watching and documenting the fabulous proliferation of scientific instruments, papers, teams, ideas and skills milling around a minute quantity of endorphin! No matter how reductionist a neuroscientist claims to be, read his or her papers: they multiply new concepts, new diagrams, new theories, new names to welcome in their world any new agency that in their terms explains all the others. Reductionism, and thus the fight against it, is a misplaced target, since the more simple are the agencies that are supposed to eliminate all the others, the bigger, vaster, costlier is the scientific assembly, instrumentation and institution that hold them in place. Such is, in my view, the interesting cyborg: the conjunction of the matters of fact with the rich vascular system of scientific production that maintains their existence. By fighting or celebrating reductionism, by opposing the phenomenological and physiological body, or worse, by claiming to reconcile both into a makeshift holistic body and soul, we lose a formidable chance for body politics to highlight the scientific activity itself.

Why is this a good idea? First, it is a chance for scientists to be freed from the impossible burden of defining a shared world made of primary qualities. Now might be the time to deliver scientists from their lonely task of defining this shared world. We should be able to benefit from the rich multiplicity of sciences, without having to pack all of them into one kingdom, reserving subjectivity for a meaningful but essentially meaningless world of arbitrary fictions. Second, it is useful because it opens up to a considerable degree the range of potentialities, since the bodies politic may now profit from the realities of the science without being immediately accused of incompetence because of the perceived sharp distinction between facts and values, primary and secondary qualities. Subjectivity is no longer what is left when the primary qualities have taken their lion's share, but what subjects us, what affects and effects us. Subjectivity does not live inside, in the cellar of the soul, but outside in the dappled world. Third, if pluralism is expanded to the sciences, they become sortable, accountable, and their multiplicities may now be added to those of political assemblies and disputes; they gain colors. This is a far more realistic vision of what the contemporary issues surrounding biopower have in store for all of us. It is on both sides of the former front line, mononaturalism and multiculturalism, that the new dividing lines can make some progress. Even if the sound of the word 'multinaturalism' is offputting at first, how productive would be its introduction into common parlance if it could overcome the limits of the impossible solution of multiculturalism: the abandonment of the shared world into incommensurable and irreconcilable multiplicities. How much more rewarding could it be to reopen *with* the sciences, not against them, the task of defining the shared world – what Isabelle Stengers has called 'cosmopolitics'.

In conclusion, let me turn to Belgian philosopher Vinciane Despret, commenting on the extraordinary sentence of William James 'Our body itself is the palmary instance of the ambiguous'. Despret proposed to define the body thus: 'that through which we learn to be affected'. The more body we have, the more finely attuned it is, the more articulate, the more skillful, the more we become sensitive to the presence and interference of others. Those who misconstrue incarnation often imagine that without a body they would roam through the cosmos with better ease; without Moravec 'wetware platform' they would download themselves through the Web with more calculating power; without instruments and artefacts, colleagues and laboratories, they would know more; without prostheses and machinery they would be freed and emancipated – soul, only soul. Those who revel in incarnation, on the contrary, want to have as many bodies as possible, to subscribe to as many sciences as possible so as to become affected by many more agencies. They know that the opposite of body is not emancipation, it is not soul, it is not spirit, it is not life, especially not eternal life; the opposite of body is death.

FURTHER READING

Despret, V., *Ces Émotions Qui Nous Fabriquent. Ethnopsychologie de l'authenticité*, Paris: Les Emp'cheurs de Penser en Rond, 1999.

Gray, C. (ed.), *The Cyborg Handbook*, London: Routledge, 1995.

Hacking, I., *Rewriting the Soul, Multiple Personalities and the Sciences of Memory*, Princeton, NJ: Princeton University Press, 1995.

Haraway, D. J., *Modest Witness@Second Millennium. Female Man meets OncoMouse: Feminism and Technoscience*, London: Routledge, 1997.

James, W., *The Principles of Psychology*, New York: Dover, 1890.

Kohler, R. E., *Lords of the Fly*. Drosophila *Genetics and the Experimental Life*, Chicago: University of Chicago Press, 1994.

Latour, B., *Pandora's Hope. Essays on the Reality of Science Studies*, Cambridge, MA: Harvard University Press, 1999.

Pickering, A., 'Cyborg history and the World War II regime', *Perspectives on Science*, 3 (1995), no. 1, 1–48.

Stengers, I., *Power and Invention*, Minneapolis: University of Minnesota Press, 1997.

## 8 The Iceman's Body – the 5000 Year Old Glacial Mummy from the Ötztal Alps

KONRAD SPINDLER

### Introduction

The following chapter is structured in four parts. The first deals with the highly serendipitous sequence of events that led to the discovery and recovery of the glacial mummy. The second focuses on the body itself and the relevant anatomical, medical and anthropological findings, and the third section is devoted to the equipment this Neolithic man carried with him on his last journey into the high mountains of the Alps. At the end there are some speculations about the Iceman's fate.

### Discovery and recovery

Between 15 and 23 September 1991, Helmut and Erika Simon, a German couple from Nuremberg, were spending a walking holiday at a little resort called Madonna di Senales in South Tyrol. From their base they went walking in the Alps every day, returning to their accommodation in the evening. On Wednesday 18 September they set off to climb Mount Similaun. On their way up the mountain they encountered problems of orientation in the difficult terrain and at one point found themselves in a dangerous crevasse zone. This forced them to make an exhausting detour, and they did not arrive at the summit until relatively late in the afternoon, 3.30 p.m., which meant it was too late for them to return to their accommodation the same day. They therefore had to find somewhere to spend the night in the mountains. The nearest hut was the Similaun Refuge; they arrived there just before dark and were forced to rest there.

The next day, Thursday, 19 September, the Simons planned to return to their car down in the valley first thing in the morning, but the weather was so beautiful up in the mountains that they decided to climb another peak, the Finailspitze, which they duly reached at midday. After spending an hour resting at the summit, they set off back down again, but they left the sign-posted trail (which, of course, is not really something one should do in the mountains). Their route took them across a gently sloping ice field, where they came to a narrow trough in the rock, partly covered with glacial ice and partly filled with meltwater (Figure 1). They had then to decide whether to negotiate the obstacle by going to the left or to the right of it.

They elected to keep to the right-hand side of the trough, and as they walked round it they noticed something sticking out of the ice. At first they thought it was an old doll, as so much of the rubbish of our affluent society is left lying around even in the high mountains. But as they moved forward for a closer look they discovered to their horror that it was a human body, with the head and shoulders protruding from the ice (Plate I). In spite of the shock, Helmut Simon had enough presence of mind to photograph the scene, using the last frame of the film in his camera.

The Simons also saw a blue ski clip lying nearby, one of those thick strips of rubber that ski mountaineers use to hold their skis together when

FIGURE 1. The rock gully on Hauslabjoch where the Iceman was found during the archaeological follow-up on 10 August 1992.

143

shouldering them for the climb up, and so they assumed that the body must have been lying in the ice for about ten or twenty years. About two metres from the head of the corpse they also noticed a kind of birch-bark container. They picked it up for a closer look and then put it down again saying, 'Funny things the birds carry up the mountains these days!'.

After about ten minutes they returned to the Similaun Refuge to report the find; the first question they put to Markus Pirpamer, the warden of the refuge, was whether anyone was missing from the hut. When he said no, they reported their discovery, which of course sparked very considerable excitement. Pirpamer made the Simons describe the location as best they could, and from what they said it was not clear whether the body was lying on the far side of the main ridge of the Alps, that is to say in South Tyrol and therefore on Italian territory, or on the north side, i.e. on Austrian territory in North Tyrol. For that reason the warden telephoned both the Italian carabinieri in Senales and the Austrian police in Sölden. The Italians, however, showed no interest, and so the find became the responsibility of the Austrians.

The official name of the site of the find itself is Hauslabjoch, which is located 3210 metres above sea level, an altitude at which the thin air causes problems of lift for helicopters.

That afternoon Markus Pirpamer set off with his Bosnian kitchen worker Blaz Kulis to take a look at the find. After a short search they found the body, saw the birch-bark container, and about five metres south of the corpse, on a rocky ledge that represents the downhill limit of the trough, they also discovered a number of other items, including a stick that was still half frozen in the ice, something resembling a kind of axe, probably an ice-axe, and various flat or round pieces of wood, which reminded Pirpamer of broken snowshoes, while Kulis thought they were the remains of a sledge. At all events, everything seemed exceedingly odd, not to say mysterious. After about half an hour they set off back to the Similaun Refuge.

In the meantime, the mountain rescue service based at Vent in the upper Ötz Valley had already initiated preliminary investigations. Someone remembered that in 1941 a music teacher from Verona by the name of Carlo Capsoni had gone missing on a walk from Belavista to the Similaun Refuge via Hauslabjoch, and that seemed to settle the question of the identity of the corpse for the time being.

By then it was too late for a helicopter to fly to the site of the find to recover the body, and so the flight was scheduled for the following day, 20 September. Because of poor weather in the morning, the helicopter did not in fact take off from Innsbruck Airport until almost midday. The pilot Hermann Steiner was accompanied by Anton Koler of the Austrian police and the mountain rescue service, while Markus Pirpamer had been asked to climb up to Hauslabjoch to show the pilot where to land. They all arrived at the site at the same time, at 1.16 p.m.

With the help of a jackhammer, or pneumatic chisel, the two men attempted to cut the body free from the ice. In the meantime the helicopter had taken off again because of the deteriorating weather conditions and an approaching storm threatening to blow the light aircraft off the mountain. With the body still enclosed in ice from the hips downwards, the compressed air supply ran out and the operation had to be discontinued. As the melt-water had continued to fill up the depression, much of the work had to be performed under water, and in some cases the chisel slipped and cut into the flesh of the corpse, causing considerable damage to the left hip and thigh.

The two men called up the helicopter to collect them, and while they were waiting they took a closer look at the find. They noted an injury to the skull, and also a number of black stripes on the back of the corpse, which they thought could be brand marks or scarring caused by a whip. That raised the question of third-party involvement, and for that reason the public prosecutor and an examining magistrate were called in, and they initiated criminal proceedings against 'persons unknown', proceedings which were later discontinued of course.

A first view was also taken of the various artefacts found on the rocky ledge. They saw the dead man's bow with the lower part still firmly encased in ice, the axe with the metal blade lying across it at an angle, and the flat and round pieces of wood, which were later identified as pieces of a backpack frame, not unlike the metal support incorporated into a modern frame rucksack. The trough itself contained pieces of leather and fur. The police officer thought the axe looked fairly old, and that suggested the nineteenth century to him as the date for some kind of mountain incident, so that the Veronese music teacher could not be the answer after all. As evidence in support of his theory he placed the axe in his bag and took it down to the valley with him. That was to have regrettable consequences as it meant that the only more or less datable item had been removed from the site of the find, with the result that the prehistoric origins of the corpse were not established until much later.

Shortly afterwards the helicopter arrived, picked up the men and flew down to Vent as the first stop. Vent is located at the very end of the Ötz Valley and is literally the end of the road for cars. After that, access is on foot only, and the climb up to Hauslabjoch takes about four and a half hours at a good pace.

At the landing site in Vent, Anton Klocker, the undertaker in Längenfeld in the Ötz Valley, had been waiting since 1 p.m., having received instructions about midday to collect a certain Mr Capsoni.

The landing site was also the scene of a lively discussion on the subject of the axe that the police officer had brought down from the mountains. Alois Pirpamer, chief of the mountain rescue service in the upper Ötz Valley, was hoping to acquire it for the mountain guides' corner in his Hotel Post, while the undertaker wanted to take it to the regional heritage museum in Längenfeld, but the police officer insisted on following the correct procedure for something that was for the moment state property and took it to the police station in Sölden.

Next day, Saturday, the mountain rescue post in Vent was instructed that recovery of the corpse had been postponed until Monday because – they were told – the helicopter had more important missions to fly over the weekend in the way of traffic control, road accidents, etc. For that reason the warden of the hut climbed up to the site of the find again in the early morning. He covered the mummy with some black plastic sheeting, which was in fact a plastic garbage bag, cut open so as to protect the corpse from the walkers expected on the scene at the weekend. On his way back to the refuge, in the rocks of the next ice-free hollow at about twenty metres distance from the mummy, he found a largish piece of birch bark. Pirpamer thought that strange at such a high altitude and therefore put it in his pocket and took it back to the Similaun Refuge with him.

In the afternoon a number of people arrived at the refuge towards 3 p.m., namely the famous South Tyrolean mountaineer Reinhold Messner, his climbing partner Hans Kammerlander, and a mountain guide by the name of Kurt Fritz, who was accompanying the other two that day on their circular tour of the mountains of South Tyrol. Waiting in the refuge were Dr Hans Haid, an ethnographer from the Ötz Valley, and his wife Gerlinde, a folk music researcher, who had arranged to meet Messner there. Of course the conversation immediately focused on the body in the ice. The warden showed the piece of birch bark he had collected that morning, and from memory he produced a sketch of the axe, which was by then in the keeping of the police in Sölden.

The party immediately decided to set off for Hauslabjoch. The three mountaineers, Messner, Kammerlander and Fritz, took a short cut up a scree slope and reached Hauslabjoch in thirty minutes. Hans Haid took forty-five minutes, and his wife – having lost her orientation in the dangerous mist – one hour. That, by the way, means that the old rule of the mountain – that the slowest member of the party should set the pace – was disregarded.

On the mountain the plastic sheet was removed, photographs taken, and a discussion begun. The ice that formed on the meltwater overnight was chipped away around the mummy. Messner used a ski pole for the work, and Kammerlander a piece of wood he found lying on the ground. That happened to be a part of the backpack, which the Man in the Ice had used to carry his possessions up to Hauslabjoch 5000 years earlier. So it was still doing good service.

In the meantime the birch-bark container had been trampled, and the contents scattered over the ice. It is particularly important to note that it was possible to see without a shadow of doubt that the dead man was wearing something resembling trousers. That has since been confirmed by witnesses. They could even make out the fine seams of the garment, and Messner also saw a shoe with hay spilling out of it, which reminded him of the footwear worn by the Lapps. This is important because it was later claimed that the body had lain naked in the ice.

In the meltwater the mountaineers noticed a small wooden stick barely twenty centimetres long. It had holes at regular intervals, which reminded the group of a flute. Fortunately, Gerlinde Haid, the authority on folk music, had arrived on the scene and was able to correct this misconception. In fact it is part of the reinforcement from the quiver. This fragment was not found near the quiver itself, which was only discovered several days later at some distance from the ice mummy.

After about an hour the group set off back to the Similaun Refuge, where Paul Hanni, Messner's manager, was waiting to be briefed on the day's events ready for a press release. Interestingly enough, Messner at that point expressed the opinion that the body was at least 500 years old, maybe even 3000. So he was already relatively close to the truth. Unfortunately the newspaper reporter who decided to run the story thought that 500 years was enough, and that led to the fairy story in the next day's papers about a mercenary of Count Frederick of Tyrol (1403–1439). Following a less than successful skirmish in Upper Italy at about that time Frederick had in fact attempted to withdraw up Val di Senales, past Hauslabjoch and down the Ötz Valley to Innsbruck, so that it was not unreasonable to suggest that one of his men might have had to be left behind in the mountains. At least that meant the Man in the Ice had now reached a ripe old age of 500 years.

That day a report from Officer Niederkofler of the Carabinieri in South Tyrol arrived to the effect that Carlo Capsoni, the music teacher from Verona, had been buried in Italy back in 1952, which ruled out that theory once and for all.

Next day, Sunday, Mountain Rescue Chief Alois Pirpamer himself climbed up to Hauslabjoch. Together with a pensioner by the name of Hans Gurschler, he intended to hack the body free from the surrounding ice in readiness for the official recovery, which was planned for the following day. They were successful in their mission and actually managed to raise the body clear of the ground. In the process they noticed various scraps of leather and fur adhering to the ice. They gathered up the various artefacts lying around and took them down to Vent. Pirpamer notified the mountain rescue service that the body was ready for collection, and once again Klocker, the undertaker, asked to be duly informed of the time of the recovery flight.

Monday, 23 September, was the day of the official recovery – four days after the initial discovery had been made. Again take-off was delayed until the weather conditions improved, and towards midday the helicopter took off from Innsbruck Airport piloted by Anton Prodinger, with police officer Roman Lukasser from the air rescue service and Professor Rainer Henn of the Department of Forensic Medicine at Innsbruck University – and therefore responsible for the Innsbruck district – as his passengers.

After a brief stop in Vent, the helicopter touched down at Hauslabjoch at 12.37 p.m., and Henn stepped out of the helicopter to be confronted by two surprises. The first was the presence of a complete camera crew from Austrian Television (ORF), who proceeded to film the whole process of recovery. As one knows, their footage was broadcast worldwide and much of what was shown led to bitter criticism in the world of forensic medicine of the methods employed to recover a prehistoric corpse.

The second surprise was the fact that the body had once more frozen solid in the ice. Unfortunately no-one had thought to bring any tools or equipment because they had been told the body was ready for collection. So the party was standing around feeling fairly useless when along came Markus Wiegele, a mountaineer from South Tyrol, and offered the recovery team the use of his ice axe and ski pole to release the body from the ice once again, with the result that the term 'ski-pole archaeology' has now been added to the technical language employed in archaeological field work.

So they set to work to hack the corpse clear of the ice. As soon as the body could be moved it was pulled to one side with the help of the ice axe and ski pole. Finally the body was lifted from its icy grave and placed at the side of the hollow. Further items were fished out of the meltwater. The forensic examiner found a largish item and dropped it on the ice, where it rebounded. It was a dagger with a stone blade and wooden handle.

That at least should have made every one sit up. But, as Professor Henn later explained, when he saw the dagger his first thought was that it was something an escaped convict had made himself.

In the meantime Wiegele made a last attempt to extract the bow from the ice, but the ice would not release its grip and the bow broke. The end left in the ice was not to be salvaged until the following year in the framework of the large-scale archaeological follow-up study of the site at Hauslabjoch, but at least the bow is now complete again.

As is normal practice, the corpse and the other finds were placed in a transparent body-bag with a zip fastener, but it was not zipped up completely as the mummy had started to smell. The body-bag was then inserted into the orange recovery bag and clipped on to the underside of the helicopter. And so, over 5000 years after his solitary death amidst the mountain peaks of the Alps, the Man in the Ice made one last descent to the valley, where it was finally the turn of the patient Mr Klocker, the undertaker.

The mummy was accordingly placed in a coffin, an act that required bending the left arm, which was projecting at an angle. That produced an audible crack, as Klocker informed. The radiologists were later to diagnose a fracture of the left humerus. The coffin lid was then screwed down. At almost

2.30 p.m. the undertaker finally drove off towards Innsbruck, stopping on the way in Sölden to pick up the axe that had been handed in to the police. At 4.06 p.m. Klocker arrived in front of the Department of Forensic Medicine in Innsbruck, and the body was handed over to Dr Unterdorfer.

The corpse and the various artefacts were placed on two tables in the dissection room. One table was used to spread out the freshly broken bow, some leather thongs which went with the belt, remains and scraps of clothing, another piece of the belt with a sewn-on pouch, and also the axe with its metal blade, while the other table was reserved for the mummy, with other remains of clothing. That was the moment the medical colleagues had a first inkling that this could in fact be a much older find. For that reason they contacted the Department of Archaeology. Even then, Konrad Spindler's initial assessment that the find was at least 4000 years old if not more met with utter disbelief. The scepticism was not unreasonable given the fact that a human body cannot normally remain in glacial ice for such a long period. Even in the case of extremely long and gently sloping glaciers, which in consequence have a very slow rate of flow, the accepted wisdom was that dwell time could not exceed 600 years. That made it imperative to invite the glaciologists to make their contribution.

They accordingly started making preparations for a first expedition to Hauslabjoch the next day. Unfortunately, while the practised mountain walkers from the Department of Glaciology made it on foot to the site of the find, the helicopters chartered for the archaeologists were again forced to turn back because of the bad weather conditions. With a full-blown *föhn* rolling in from the south and a bank of mist building up, it would have been irresponsible to try to land at Hauslabjoch. Down on the ground, however, a member of the walking party, Heralt Schneider, discovered the quiver the minute he arrived at the site. It was frozen solid to the bottom of the rocky depression and surrounded by glacial ice.

Perhaps this is a suitable moment to discuss very briefly why the rate of glacial retreat should have been so unusually high at that time. As one knows, winter precipitation levels are the key factor for the thickness of the glacial ice cover. In simple terms, the glaciers grow when more snow falls in winter than melts in summer, and vice versa. In the three years prior to 1991, precipitation in the Alps had been relatively strong and the rate of retreat observed in glaciers everywhere had been correspondingly pronounced.

The Pasterze, for example, the biggest glacier in the Eastern Alps located at the foot of the Grossglockner, retreated by about 170 metres in those three years. The crucial factor in 1991, however, was a desert storm in the Sahara, with winds that deposited a dirty brownish-grey layer on the ice fields. As a result, the sunlight was not reflected by the white snowfields, as is normally the case, but was absorbed by the dark layer of dust, which warmed up accordingly and produced extraordinary rates of glacial melt, peaking at about ten centimetres loss of magnitude a day in the second half of September. Ultimately, then, we owe the emergence of the Iceman in 1991 to a desert storm in North Africa.

To return to the events on the mountain, the glaciologists made the only correct decision when they heard the helicopter with the archaeologists turn back; they recovered the quiver. And they did so without even once touching it with a tool, by continually pouring meltwater over it, a process that took about two hours. Just as the quiver had been extracted from the ice and the glaciologists were wrapping it up in quilted jackets and splinting the bundle with ski poles, a dark figure suddenly emerged from the mist. It was an officer of the Italian border guards by the name of Vice-Brigadier Silvano dal Ben. Fortunately one of the members of the group was able to keep him talking while the others hastily stowed the quiver in a rucksack and vanished into the mist themselves.

That proved to be a wise move, as the officer subsequently removed articles from the site, above all some of the pieces of the Iceman's clothing, and took them to Bolzano in Italian South Tyrol. As a result, a battle of red tape had to be fought for no less than nine months before these items could be re-united with the rest of the Iceman's equipment at the Roman Germanic Central Museum in Mainz where it had been sent for restoration in the meantime.

A first archaeological examination of the site at Hauslabjoch was not possible until weather conditions finally improved on 3 October. By then sixty centimetres of fresh snow had fallen, which greatly hampered the work. Nevertheless a number of further items were recovered and initial surveying performed using a tacheometer so as to have an exact record of the location of the site for the resumption of work the following year. That was a good thing, too, as the 1991/92 winter brought seven metres of snow to Hauslabjoch, and the snow cover did not melt completely the following summer. Two days later, on 5 October, the onset of the first winter storms put an end to field work for that year.

### Back in the laboratory

The mummy is in an excellent state of preservation; it is doubtless one of the best preserved human mummies in the world. The body dehydrated in the glacial ice, a process with which every housewife is familiar who has ever placed a piece of unwrapped meat in the freezer (Figure 2). As a result of incipient epidermal decay, the body is completely hairless, but numerous hairs were found among the various artefacts indicating that the Iceman had had dark brown to blackish, slightly wavy, shoulder-length hair and also a beard.

The very first scientific investigations we performed, in the evening of the day on which the prehistoric character of the find had been established, were based on computed tomography (CT) scans and X-rays. It was necessary to create full records of the mummy as it was far from certain that it would not crumble into a pile of dust by the following day. The images produced indicated that the internal organs were well preserved. The brain with the

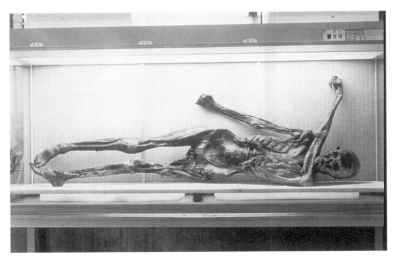

FIGURE 2. Anterior view of the Iceman in a micro-flow box at Innsbruck University's Department of Anatomy.

various cerebral convolutions had shrunk to about two-thirds the original size. Conservation is so good that the fine meninges are fully preserved.

The CT data were then employed to build a three-dimensional skull model using stereolithography. This was a world first for a technique that met with great interest in medical and also anthropological circles at the time and is now a routine procedure, especially for reconstructive surgery for cranial deformations.

Investigations of the internal organs were subject to considerable delays at first. This was because of the need to develop suitable tools as conventional equipment was not sterile enough. An American company then developed titanium alloy instruments, which are designed to avoid all contamination. They provided the first view of the internal organs of the body. The lungs are blackened, as in the case of a chain-smoker today. That is due to the fact that the Iceman spent long hours sitting in front of an open fire and breathed in the smoke in the process. It was also able to take a first look at the Iceman's heart. That is also shrunk, but a coronary vessel is clearly visible nevertheless.

The Iceman's face, with its harmonious features, illustrates once again the excellent state of preservation of the mummy. Conservation also extends to the eyes, complete with cornea, pupil and iris, and one almost feels one can identify their colour. Only in the middle part of the face are the soft parts distorted, with the nose and upper lip displaced upwards and to the right. The explanation is that the man died in his sleep lying on his left side and was covered with snow during the night. Were that not the case the body would have been discovered by carrion eaters, which in the high mountains means primarily vultures, eagles and crows. In fact the mummy shows no signs of such attack.

In the following decades and centuries a slight deterioration in the climate occurred. The trough with the body lying at the bottom filled with ice, the glaciers advanced and the ice masses started to flow. The ice in the trough was held back by the rock wall but the main body of the glacier flowing over the depression also created a degree of creep in the basal ice, with the result that the body was rotated through 90° from the original position on its left side to a prone position. That rotatory movement also pressed the upper arm into an unnatural position up against the chin and, instead of resting on one cheek and temple, the head was left with the face pressing against the rough surface of the rock, which caused the flattening of the middle part of the face.

These movements also exposed the Iceman's teeth, especially the upper jaw, and here, too, some interesting details were produced in the very first anthropological investigations. As is to be expected in Neolithic man, the teeth are heavily abraded. They look longer than they really were because shrinkage of the gums has exposed part of the roots. The Iceman's teeth are completely free from dental caries – much to our envy today – and only slight plaque formation can be detected. The most striking feature of course is the gap between the two first incisors. This is what in modern dentistry is known as a congenital trema. That is a variant of normal human dentition that is still to be found today in about 10% of the older population, while younger people are more likely to have had their trema reduced by the art of the dentist.

On the first examination at the Department of Forensic Medicine, the sexual organs were described as desiccated and foliate in shape and presumably male. A more precise anatomical examination showed that the penis was completely preserved with a sickle-shaped fold representing the foreskin, and the scrotum can also be identified. Only at the basal margin of the scrotum are certain lesions to be seen, which were inflicted during recovery from the glacial ice. Unfortunately these observations were seized upon by the media and a completely unfounded castration theory was developed. Even such a reputable German publication as *Der Spiegel* printed an article claiming that the Iceman's testicles had been scraped out and that the penis was missing. There is no truth in any of that.

As the body was almost completely naked except for the right shoe on arrival at the Department of Forensic Medicine, a rumour developed to the effect that the Iceman had lain naked in the ice. That is not true, either. The shoe had to be removed for conservation purposes, all the more so as the body and the various artefacts were to be preserved at different locations (Figure 3).

In the first reports from the site at the time of the initial find, mention was made of marks on the dead man's skin. They include two parallel stripes around the left wrist, and larger areas of tattoos on both sides of the lumbar vertebral column, which can be seen clearly through the skin. The tattoos continue further down but have not survived the damage inflicted by the

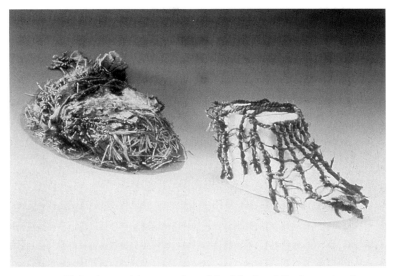

FIGURE 3. Right shoe and inner netting of the left shoe following restoration.

jackhammer during the first attempt to recover the body. One can also detect markings on the Iceman's knee in the form of a cross, and one can see very well from the bluish colour that charcoal was used for the pigment. The technique involved pulverising the charcoal and adding saliva or warm water to make a paste, which was then applied via fine incisions or punctures in the skin. When such wounds heal, the typical bluish tattoo marks remain that are still to be seen in certain marginal groups of society today. When the shoe was removed, more tattoos were found and laboratory tests confirmed that charcoal had in fact been used. A tiny particle of skin was excised for the relevant histological investigations.

That raises the question of the significance of the tattoos. In both primitive and civilised societies we know of a whole host of reasons for the practice. In this case, however, the scientists had a theory from the very start based on the observation that the tattoos were all located on joints that are subject to pronounced loading, namely the wrist, lumbar vertebral column, and the knee and ankle joints. They were therefore very curious to see the first X-ray pictures, and in fact they showed that all the joints involved displayed discrete to medium arthritic changes that are typical of the age-related attrition of the joints that is associated with such disorders as gout or rheumatism. The joints would of course give pain to the Iceman, and that is why he had his skin tattooed in these areas. Among primitive peoples, such as the Berbers in North Africa, it is an established practice to have tattoos applied to painful parts of the body. Therapeutic cauterisation, that is the practice of burning herbs into the skin over painful joints, leaving cicatricial tattoos, is also found in Tibetan medicine. The Chinese practice of acupuncture is a similar homeopathic procedure. It must therefore be concluded – with all due respect – that therapeutic practices were employed in Europe over 5000 years ago which have retained their significance down to the present day, notwithstanding the probability of a certain placebo effect.

Finally, there is recent evidence of a wound inflicted by an arrow. In 1992, only one year after the discovery of the body, preliminary X-ray photographs had shown an unidentified shadow in the area of the left shoulder blade. Advances in imaging processes since then have revealed the shadow to be caused by a 27 millimetre long flint arrowhead. The entry path is five to eight centimetres in length and the wound shows no traces of any healing process. Present research is concentrating on possible damage to nerves and blood vessels but there is evidence of bleeding between the shoulder blade and part of the rib cage. Very probably this wound was not immediately lethal.

### Clothing and equipment

Most of the garments were made of fur, but, as in the case of the body itself, the skins have lost their hairs, so that the various items now appear to be made of leather. On the other hand, this enables us to appreciate the accuracy with which the seams have been sewn, and indeed this led the forensic examiner to exclaim on initial inspection, 'Sewing machine!'. The main sewing materials used were made from animal sinew fibres. Apart from that there are some mended seams stitched together with a woollen thread and also a number of clumsy repairs where the holes have simply been pulled together with a twisted grass thread. Clearly, more than one pair of hands had been at work on the garments, namely the skilful and experienced hand of the person who originally fashioned the garments and kept them in good shape, and the clumsier hand of the Iceman himself, who would be forced to make his own repairs if he tore a garment while out on his own away from his base. He was not therefore the outsider or outcast he has so often been described as, forced to live out his life and die alone in the mountains.

Needless to say the scientists studied the tufts of fur from the garments and made some very interesting little discoveries, including two grains of cereal. These were not part of the Iceman's provisions, however, but rather grains that had been caught in his clothing by chance, trapped in the fur, and then carried up to Hauslabjoch unintentionally. The observation is all the more important as the grains had been freshly threshed. That can be seen from the sharp fracture lines and the fact that the grains were still in their husks. Botanical analysis has revealed that the grains are a very early type of cultivated cereal which was an important nutritional factor in the Late Stone Age. This find also shows that, shortly before his death, the Iceman must have spent some time in a human settlement where the harvest had been brought in and the crop threshed before he set off for the mountains.

The shoes are a particularly ingenious design. Above all, they are the oldest shoes ever found in Europe. Indeed, this is the first ever find of a complete set of Neolithic garments. Hitherto the oldest finds of European clothing had dated back only as far as the Bronze Age. The discovery of the glacier mummy has therefore extended our knowledge of European costume history back into the past by more than 1500 years.

The only garment that is made of a textile material is the grass cloak, which was tied from stalks of alpine sweet grasses over a metre in length.

An especially exciting find made during the second archaeological expedition to the site, one year after the original discovery of the mummy, was the Iceman's cap (Figure 4). It was found at the base of the large boulder on which the corpse was resting. The cap is complete with a chin-strap, which was knotted together under the chin, and it is apparent that the chin-strap had been torn during the Iceman's lifetime. This suggests that he had been dragging himself forward during the last few minutes of his life until he fell and failed to muster the strength to stand up again. In the process he lost his fur hat, which was no longer held in place by the chin-strap. The man therefore lay bare-headed in the icy night of the high mountains and was doomed to freeze to death.

The Iceman also wore a belt made of calfskin, with a pouch sewn onto the belt at the front. The belt was also used to attach a pair of leggings not unlike those worn by North American Indians. They are made of goatskin

FIGURE 4. Cap made of bearskin with chin strap.

– from the skin of the domestic goat, to be precise. That again shows that the man must have had something to do with domestic animals. At the lower end the leggings have a kind of tongue that was inserted into the shoes and tied in to keep the cold out. At the top there are straps to fasten them to the belt.

The belt was also used to hold the loincloth in place. The latter is made of a very fine and thin piece of goat leather and was worn hanging down to the knees at the front so as to protect the contents of the belt pouch as well.

The outer garments worn by the Iceman include a coat. This is again made of goatskin, but this time the furrier carefully selected strips of fur of different colours so as to create a patterned effect based on light and dark vertical stripes. Originally the coat must have been a rather splendid item.

On top of the coat the Iceman wore a grass cloak, which is made of grass divided into sections with lengths of twine knotted in horizontally in the upper part, with a long loose-hanging fringe around the bottom providing adequate freedom of movement.

The equipment includes two small larchwood boards, which are notched at both ends, and a U-shaped hazel switch with matching notches measuring over two metres in length. The switch is in several pieces but is otherwise complete. Microscopic examination of the lower part of the two ends revealed impressions left by pieces of cord. It can safely be assumed therefore that the whole thing was lashed together to form a kind of backpack of the frame type.

Interestingly enough, the cords are made of grass. That may not sound very functional by modern standards but let us not forget that in the period of austerity following the Second World War string was even made from twisted paper, and that was strong enough for many purposes, too. So this again is a technology that was developed over 5000 years ago and has not had to be changed fundamentally or improved down to the present time.

Strings were also used to make the net that was found. Although not very well preserved, it is clearly a net with a relatively wide mesh – too wide for fishing, for example. The alternative would therefore seem to be a bird net.

The only item found among the Iceman's equipment that can perhaps be classified as ornamental is a tassel, which has a dolomite bead attached to one end. The bead must have been an amulet or had some similar apotropaic or magical function designed to ward off evil. In the case of the tassel itself, however, which is made of twisted leather thongs, it is also possible that they were the Iceman's store of replacement materials. After all, the Iceman's clothing and equipment incorporated leather thongs for a number of items, including the axe and quiver. With the tassel, therefore, the Iceman always had an adequate store of repair materials with him, which again suggests that his equipment was designed to sustain a way of life that involved long periods of absence from human settlement and reliable sources of supply.

The two birch-bark containers found among the Iceman's possessions each had a capacity of about two litres. In one case it is not known what the contents were as the results of micro-analysis of the internal walls are not yet available, but it is clear that the second container was used to transport embers. That is a well-known practice among primitive peoples, hunters and nomads. When camp is struck in the morning they take a handful of embers, wrap them in damp leaves or grass, and place the package in a basket, earthenware vessel, leather pouch or a bark container of the type carried by the Iceman. That keeps the embers burning throughout the day so that a camp fire can be quickly lit in the evening.

Leaves – of the Norway maple – were in fact discovered at Hauslabjoch. They were found to contain chlorophyll still, which shows that they had been freshly picked. One can see how the leaves had been stripped off the tree, taking just the flat blades and leaving the tougher stalks. The Iceman stripped them off a branch and wrapped the embers in them. Numerous charcoal flakes and particles were detected on the leaves, and here too it has been possible to identify the types of wood used. In fact, the species involved include trees that grow down in the valleys (alder and elm), trees that are found in the high forests (spruce and pine), and also smallwood or dwarf shrubs of the willow family.

This shows that the Iceman had gradually climbed from the valley floor to higher and higher altitudes in the mountains and that he had occasion-ally lit camp fires. These are further interesting clues to events that took place in the weeks and days prior to his death.

One item of the Iceman's equipment that was originally something of a mystery looks like the stub of a pencil. It is made from a piece of a branch of the lime tree and has a pointed end with a small lump of harder material protruding. At first the scientists thought it must have been used to strike sparks. In the X-ray photographs, however, they could see that the hard lump was in fact a spike extending into the shaft. Examination of the material finally indicated that the spike was made of a splinter from a stag's antler and that the object was in fact a retoucheur; that is, a tool used to rework or sharpen flint implements. Anyone who used flint tools needed such an implement. Nothing like that had been found before, however, and the idea that something of the sort must have existed had hitherto been the subject of hypothesis only. Now we have the first authentic specimen!

The scientists were also somewhat mystified by two other items. After a whole series of investigations culminating in chromatographic analysis, the mycologists concluded that they were pieces cut from the fruiting-body tissue of the birch fungus (*Piptoporus betulinus*). The initial assumption had been that the material was tinder. That would have fitted in well with the fire-lighter. But that could not be the case, as birch fungus is unsuitable for the purpose as the tissues are incapable of glowing combustion.

On the other hand, there are writings on folk medicine suggesting that the birch fungus was offered by pharmacists as an antibiotic staunching agent right up to the twentieth century. Those who know their Alexander Solzhenitsyn will perhaps remember the scene in his moving book *Cancer Ward* where the patients discuss the birch fungus, which Russian people used to collect and dry so as to produce a fine powder for making infusions. The brew was then drunk as a homeopathic remedy for cancer based on the fact that the birch fungus was like a cancerous growth on dead and diseased birch trees. To that extent the Iceman with his two pieces of fungus threaded onto a leather thong was carrying a kind of travelling medical kit with him.

The Iceman's weapons are a varied collection, starting with a longbow measuring 1.82 metres in length (Figure 5). That is a weapon still much esteemed by modern archers. Today a good bow made of yew costs about $(US) 2000 or more. On initial inspection it could be seen that the bow was unfinished. Neither the thick grip in the middle nor the flat tension sections had been shaped, and it was lacking the notch required to take the bowstring. It can be assumed therefore that the Iceman had lost or broken his old bow shortly before his death and was forced to replace this – literally – vital piece of equipment as quickly as possible.

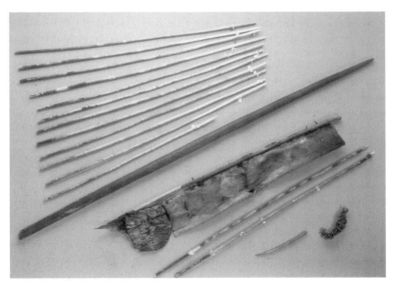

FIGURE 5. The Iceman's bow, arrows and quiver.

The quiver consists of a leather sack made of the skin of the chamois. Along the side seam the quiver is reinforced with a hazel rod, which is unfortunately broken in several places. It is quite clear that the quiver had already suffered significant damage in prehistoric times. The middle piece of the stiffening had broken away, and the Iceman was carrying it separately. He no doubt intended to repair the quiver at the earliest opportunity; only death prevented him from doing so.

The points raised about the quiver are also valid for its contents. It contains only two arrows primed for use, and they too had already been broken in several pieces during the Iceman's lifetime. The damage had been done with such violence that in one case even the stone tip of the arrow was fractured. Apart from that the arrows are fine pieces of workmanship, with shafts made of viburnum shoots, heads of leaf-shaped flint worked on both surfaces with a retoucheur, and the whole thing bonded with birch tar reinforced with threads made of animal sinew. The forward section of the shaft is slightly thicker than the rest. That ensures good stability in flight. One of the two arrows is a composite design, i.e. the shaft is made of two parts, probably a repair job. In that case the short front section of the shaft is not made of viburnum like all the others but of cornel wood, a tree that grows only south of the main Alpine chain. In addition to other clues, this suggests that the Iceman's home and the area of his wanderings lay to the south of Hauslabjoch and the Ötztal Alps; that is in Val Venosta, the upper Val d'Adige in South Tyrol.

The same meticulous care had been lavished on the tail ends of the shafts. They were first grooved and the hollows filled with birch tar to take the three-fold radial fletching, which is set at a slight angle so as to impart spin during flight, and the whole thing was wrapped round with an extremely fine thread made of nettle fibres just 0.15 mm thick. There was another fascinating discovery made. In one case the thread is wrapped clockwise and in the other counter-clockwise. That can only mean that one arrow was made by someone who was left-handed and the other by someone who was right-handed. In other words the Iceman can have made only one arrow himself at most; he must have obtained the other arrow somewhere else. Of course, one can only speculate about the possible origins of the second arrow.

The other arrow shafts found in the quiver were unfinished. They were all made of viburnum wood, over eighty centimetres long, which had been stripped of bark and notched at the thicker end to take the arrow heads. This suggests that the Iceman had shot most of his arrows shortly before his death. The last two primed arrows had been broken as well, so that he was again forced to replenish his supply of arrows as quickly as possible. All this is indicative of a situation in which the man was forced to act under very great pressure and in a state of extreme stress.

What else was in the quiver? It contained an approximately two metre length of string. Unlike all the other twisted cords found at the site, it was not made of grass but of a very tough tree bast. It was probably intended for use as a bowstring.

The quiver was also found to contain a bundle of four antler tips tied together with a length of bast. This would seem to have been the raw material for making new arrow heads. The pieces would have sufficed to carve about twelve to sixteen heads.

Another item in the quiver was a spike, a tool with an angular grip and a sharply fashioned point. In subsequent investigations it was revealed that the spike was covered in blood. Probably it was a tool for gutting and skinning animals.

Finally, there was also a mysterious bundle in the quiver. The zoologists identified the bundle as comprising two Achilles tendons freshly cut from an animal the size of a stag or cow. When such material is dried and the individual fibres separated it is possible to twist a yarn that is the equal of a modern nylon thread for breaking strength. So this was again part of the Iceman's collection of replacement and repair materials.

The contents of the calf-leather pouch included three flint implements: a scraper, a drill, and a tiny but razor-sharp blade. Traces of down found on the blade show that it had been used for shaping the fletching for the arrows. The pouch also contained a bone awl with a very fine point that had suffered secondary deformation. Whether the damage was inflicted in prehistoric times or during recovery cannot be said. The needle-sharp point is suitable for working furs or leather and even for tattooing.

The contents of the pouch also included a black substance, which at first was very difficult to identify. It was eventually shown, however, that this was a piece of true tinder fungus (*Fomes fomentarius*). Adhering to the tinder are fine pyrites crystals, indicating that the Iceman was not only carrying live embers with him but also had a firelighter – a belt and braces approach, as it were. At all events his equipment represented a wise collection of highly practicable items.

The blade of the dagger is made of flint, again retouched on both sides, but it had also been broken in prehistoric times (Figure 6). The point is missing and, although the Iceman had a retoucheur with him, he had not yet repaired the damage. Nor would he ever do so. The grip is made of ash, a wood that is still often used for the shafts and handles of various utensils today.

The sheath is plaited from fine strips of bast and is an excellent piece of workmanship. Here too it is not unreasonable to suspect that the delicate work was not performed by the coarse hands of the Iceman, but that somewhere he had gentler hands to assist him.

In addition to many other research projects devoted to various problems raised by the discovery of the Iceman, there is also the basic question of

FIGURE 6. Dagger and scabbard.

where the raw materials came from that were used to make his equipment, for example the flint. For that reason geological investigations were conducted over a wide radius of the Alps and well into northern Italy, and the geologists pulled off quite a coup in terms of topographic research, discovering the very mine where the flint that the Iceman had used to make various utensils was extracted over 5000 years ago. It is located in the Monti Lessini, about 120 kilometres south of Hauslabjoch, roughly east of Lake Garda.

A comparison of the material mined in the prehistoric period in the Monti Lessini with the properties of the flint used by the Iceman shows that they are clearly identical.

The dagger, too, which is made of a somewhat lighter variety of flint, can be matched exactly with raw materials from the Neolithic flint mine in the Monti Lessini. Any material that was considered very valuable in those days was also traded over considerable distances. And it is now known that exactly the same variety of flint has also been found north of the Alps in Neolithic settlements located over 350 kilometres from the origins of the raw material. This shows that trading over a distance of well over hundreds of kilometres was already a part of prehistoric life.

The axe is doubtless one of the most valuable of the finds in terms of the history of civilisation. The shaft is again made of yew, which is excellent wood, with the stump of a branch forming a natural right angle fashioned to produce the shaft. The metal blade was inserted into a slit cut in the head of the axe and wrapped with wet strips of animal hide, which then dried and contracted so that the blade was held in a vice-like grip.

The blade itself is made of copper, with only very slight impurities in the form of arsenic and silver. This shows that the blade was cast from Alpine ores.

### How did he die?

In conclusion I will say a few words about the Iceman's fate, about the circumstances that might have caused him to take to the mountains. It was noticed that, quite apart from the tumultuous events surrounding the recovery of the mummy, much of the Iceman's equipment had either been badly damaged in prehistoric times or was in an unfinished state. The man had obviously been involved in some form of violent dispute in which part of his equipment was lost or broken in some way or other.

This leads to the question of the timing of these events, which is relatively easy to answer. Among the Iceman's possessions was a sloe, a berry that ripens at the end of September or beginning of October in the valleys of the Alps. That is the time of year in which summer gives way to winter, the time of the first winter storms. It is certainly a time of the year to avoid the high mountains, where one can be overtaken by a sudden snow storm – with fatal results.

And there is the other point. The man was hurt. Obviously a man with such a relatively serious injury would not have taken off to the mountains voluntarily. He must have been forced to do so by his pursuers.

But why did he choose the high mountain region? Here again, we can enlist the help of the botanists to answer the question. The upper Ötz Valley is the location of extensive areas of high mountain pasture, which are still used by farmers from South Tyrol to this very day, in a system of migratory stock farming known as transhumance. In spring, the herds are driven up Val di Senales and past Hauslabjoch to the pastures of the upper Ötz Valley, returning to the south by the same route in the autumn. If the farmers are unlucky with their weather predictions they may find themselves and their sheep snowed in by an early winter storm.

Combining these observations to produce a logical picture, the archaeologists felt that if these grazings were in use in prehistoric times it should be possible to find archaeological traces today. For that reason they conducted surveys in the upper Ötz Valley, in the area of the extensive mountain pastures that are still used by South Tyrolean farmers today, and beneath a huge fallen rock they did in fact make an interesting discovery. The boulder came to rest in such a way that one of the lower edges forms a canopy, offering excellent shelter and protection from the elements. The first excavations revealed a number of flint tools that could date back to the time of the Iceman or older, including a blade made of the same flint from the Monti Lessini that was used for the Iceman's equipment. Other utensils were also found, but in these cases the origins of the flint have not yet been ascertained.

And that completes the picture. In summary, taking account of the various archaeological, medical and other scientific findings, one can draw the following conclusions. The Iceman was primarily a shepherd. As such he naturally carried the weapons he needed to hunt for food and protect his flocks. He spent the summer on the high pastures of the upper Ötz Valley, assembled his flock in autumn and drove them back down to the valley to his village, where the grain harvest was in full spate. But then something untoward happened – a disaster, a violent conflict, which forced him to flee. He chose the route he was familiar with in summer from his migratory shepherd life. He withdrew as far as he could, high up into the mountains; he got as far as Hauslabjoch, where he was overtaken by a winter storm and sought what shelter he could in a trough in the rock on Hauslabjoch.

A sudden break in the weather in the Alps can be fatal. In 1994, 136 people lost their lives in the Tyrolean Alps alone – in spite of modern rescue systems, transceivers, helicopters and so on – people who had underestimated the dangers of the high mountains.

FURTHER READING

Barfield, L., 'The Iceman reviewed', *Antiquity*, **68** (1994), No. 258, 10–26.

Egg, M. and Spindler, K., 'Die Gletschermumie vom Ende der Steinzeit aus den Ötztaler Alpen – Vorbericht', *Jahrbuch des Römisch-Germanischen Zentralmuseum*, **39** (1992/1993), 3–113.

Höpfel, F., Platzer, W. and Spindler, K. (eds.) *Der Mann im Eis*, Band 1, Bericht über das Internationale Symposium in Innsbruck 1992; Veröffentlichungen der Universität Innsbruck no. 187, Innsbruck, 1992.

zur Nedden, D. and Wicke, K., 'The Similaun Mummy as observed from the viewpoint of radiological and CT data', in *The Iceman*, Report on the 1992 International Symposium in Innsbruck, vol. 1, ed. F. Höpfel, W. Platzer and K. Spindler, pp. 3–19, reprint from University of Innsbruck Publication no. 187, Innsbruck, 1992.

Seidler, H., Bernhard, W., Teschlemicola, M., Platzer, W., Nedden, D. Z., Henn, R., Oberhauser, A. and Sjovold, T., 'Some anthropological aspects of the prehistoric Tyrolean Ice Man', *Science*, **258** (1992), 455–457.

Sjøvold, T., 'The Stone Age Iceman from the Alps: the find and the current status of investigation', *Evolutionary Anthropology* **1** (1992), No. 4, 117–124.

Spindler, K., *Der Mann im Eis. Die Ötztaler Mumie verrät die Geheimnisse der Steinzeit*, Munich: C. Bertelsmann-Verlag, 1993.

Spindler, K., Ötzi de gletsjerman, Het enige geautoriseerde verslag over de ontdekking van 'de man in het ijs', Al Weert: M&P Uitgeverig, 1993.

Spindler, K., *The Man in the Ice. The Preserved Body of a Neolithic Man Reveals the Secrets of the Stone Age*, London: Weidenfeld and Nicolson, 1994.

Spindler, K., 'L'homme du glacier – une momie du glacier du Hauslabjoch vieille de 5000 ans dans les Alpes de l'Ötztal', *L'Anthropologie*, **99** (1995), 104–114.

Spindler, K., *El hombre de los Hielos, El Hallazgo que releva los Secretos de la Edad de Pedra*, Barcelona: Circulo de Lectores, 1995.

Spindler, K., *Der Mann im Eis, Neue sensationelle Erkenntnisse über die Mumie aus den Ötztaler Alpen*, Munich: Wilhelm Goldmann Verlag, 1995.

Spindler, K., *O homen no gelo, Os segredos da Pré-História revelados pela descoberta de uma múmia com mais de 5000 anos*, Mem Martins: Editorial Inquérito, 1996.

Spindler, K., *Mannen i isen, En mumie fra Ötztalsalpene avslører steinalderens hemmeligheter*, Oslo: Universitets forlaget, 1996.

Spindler, K., *Mannen i isen*, Stockholm: Universitets forlaget, 1996.

Spindler, K., 'L'homme gelé. Une momie de 5000 ans dans un glacier des Alpes de l'Ötztal'. *Dossiers d'Archéologie*, **224** (1997), 8–27.

Spindler, K., *Mužz Ledove*, Praha: Edice Columbus, 1998.

Spindler, K., *Mees Jääs*, Olion Talliun, 1999.

Spindler, K., Rastbichler, E., Wilfing, H., zur Nedden, D. and Nothdurfter, H. (eds.), *Der Mann im Eis, Neue Funde und Ergebnisse*, Veröffentlichungen des Forschungsinstitutes für Alpine Vorzeit der Universität Innsbruck, vol. 2, Wien, New York: Springer-Verlag, 1995.

Spindler, K., Wilfing, H., Rastbichler-Zissernig, E., zur Nedden, D. and Nothdurfter, H. (eds.), *Human Mummies, The Man in the Ice*, Veröffentlichungen des Forschungsinstituts für Alpine Vorzeit der Universität Innsbruck, vol. 3, Wien, New York: Springer-Verlag, 1996.

# Notes on Contributors

**David Canter** is Professor of Psychology at the University of Liverpool, where he directs the Centre for Investigative Psychology. He has published widely on many aspects of social and environmental psychology, as well as acting as a consultant to major industries, government enquiries and police investigations all over the world. He is probably best known for his work on 'offender profiling' described in his book *Criminal Shadows* (1995).

**Peter Goodfellow** has worked for many years as a research scientist specialising in human genetics. His first independent position was at the Imperial Cancer Research Fund in London, where he worked for thirteen years studying human gene mapping and the genetics of sex determination. In 1992, he was elected to the position of Balfour Professor of Genetics at Cambridge University and he is currently Senior Vice-President of Discovery Research at the pharmaceutical company GlaxoSmithKline. Amongst other honours, Dr Goodfellow is a Fellow of the Royal Society and has been awarded the Louis Jeantet Prize for Medicine (with Dr R. Lovell-Badge), the Francis Amory Prize (with Dr R. Lovell-Badge) and the Soham Grammer School Prize for English Literature.

**Ian Hodder** teaches archaeology in the Department of Cultural and Social Anthropology and at the Archaeology Centre in Stanford University. Until recently he was Professor of Archaeology at Cambridge. He is a Fellow of the British Academy and has written extensively on archaeological theory and prehistoric Europe. He currently excavates in Turkey.

**Thomas Laqueur** is Professor of History at the University of California, Berkeley. He is author of *Religion and Respectability: Sunday Schools and*

*Working Class Culture 1790–1850* (1976), *Making Sex: Body and Gender from the Greeks to Freud* (1990), and *Onan's Fate: Solitary Sex and the Modern Self* (2002). The issues mentioned in the chapter in *The Body* will be discussed, among others, more fully in a forthcoming book, *Death in our Times*. Professor Laqueur writes regularly for the *London Review of Books* and the *New Republic*.

**Bruno Latour** trained as a philosopher and an anthropologist. After field studies in Africa and California he specialised in the analysis of scientists and engineers at work. In addition to work in philosophy, history, sociology and anthropology of science, he has collaborated in many studies in science policy and research management. He has written *Laboratory Life: The Construction of Scientific Facts* (1979), *Science in Action* (1987) and *The Pasteurization of France* (1988). He also published a field study on an automatic subway system, *Aramis, or The Love of Technology* (1996), an essay on symmetric anthropology, *We Have Never Been Modern* (1993), and *Pandora's Hope; Essays in the Reality of Science Studies* (1999). In a series of new books in French (to be translated into English) he is exploring the consequences of science studies on different traditional topics of the social sciences (*Sur le Culte Moderne des Dieux Faitiches* (1996) and *Paris Ville Invisible* (1998), a photographic essay on the technical aspects of the city of Paris). He is presently doing field work on one of the French supreme Courts and has recently published *Politiques de la Nature, Comment Faire Entrer les Sciences en Démocratie* (2000). He is Professor at the Centre de Sociologie de l'Innovation at the École Nationale Supérieure des Mines in Paris and Visiting Professor at the London School of Economics.

**Griselda Pollock** is Professor of Social and Critical Histories of Art and Director of the AHRB Centre for Cultural Analysis, Theory and History (CentreCATH) at the University of Leeds. She is a leading feminist thinker and art historian who helped to open the field of feminist analysis of the visual arts with the books *Old Mistresses: Women, Art and Ideology* (1981) and *Vision and Difference* (1988). Author of over eighteen books on art history, feminist theory and cultural analysis, and contemporary art, her most recent publications include *Differencing the Canon: Feminist Desire and the Writing of Art's Histories* (1999) and *Looking Back to the Future: Essays on Life, Art and Death* (2000).

**Konrad Spindler** is an expert in medicine, anthropology and archaeology, as well as prehistory and early history. He was head of excavation at Magdalenburg in the Black Forest and from 1977 to 1988 held the position of Professor at the University of Erlangen-Nuremburg. Since 1988 he has occupied the Chair of Pre- and Protohistory at Innsbruck University and served as head of the Department of Medieval and Postmedieval Archaeology. He is the author of numerous books on archaeology.

**Sean Sweeney** studied at Cambridge where he completed his Ph.D. in genetics and was a Research Fellow of Darwin College. Using fruit flies as a model organism, he studies the development and growth of the neuromuscular synapse and is currently a Wellcome Trust Prize Travelling Fellow at the Department of Biochemistry and Biophysics, University of California, San Francisco.

**Richard Twyman** studied as a PhD student at the University of Warwick and worked as a postdoctoral fellow at the Laboratory of Molecular Biology in Cambridge before becoming a full-time scientific writer based at the John Innes Centre in Norwich. He is the author of several molecular biology and developmental biology text books, including *Advanced Molecular Biology* (1998), *Instant Notes: Developmental Biology* (2000), *Principles of Gene Manipulation* (with S. B. Primrose and R. W. Old, 2001) and the forthcoming *Principles of Genome Analysis and Genomics* (with S. B. Primrose).

**Mary Warnock** was educated at St Swithun's School, Winchester, and Lady Margaret Hall, Oxford. She taught philosophy at Oxford University for fifteen years, was Headmistress of Oxford High School for six years, and Mistress of Girton College, Cambridge, from 1985 to 1991. She chaired the Government Committee of Inquiry into Human Fertilisation and Embryology, which reported in 1984. She is an independent life peer.

# Acknowledgements

Cover illustration: Imprint, 1961 (spray paint on enamel) by Yves Klein (1928–62) Private Collection/Bridgeman Art Library. © ADAGP, Paris and DACS, London 2002.

CHAPTER 1

FIGURE 5: Courtesy of Dr James Drummond, Department of Genetics, University of Cambridge.

FIGURE 6: Courtesy of Dr Bill Wisden and Alison Jones, MRC Laboratory of Molecular Biology, Cambridge.

PLATE II: Courtesy of Dr James Drummond, Department of Genetics, University of Cambridge.

CHAPTER 2

FIGURE 1: Pitkin Unichrome.

FIGURE 2: A. Barrington Brown/Science Photo Library.

FIGURE 3: The Wellcome Trust Medical Photographic Library.

CHAPTER 6

FIGURE 1: Courtesy of Mona Hatoum.

FIGURE 2: Gemäldegalerie, Staatliche Kunstsammlungen, Dresden.

FIGURE 3: © Paris, Musée Rodin (photograph Adam Rzepka) 5.1148.

FIGURE 4: Photograph copyright © 2000 New York Whitney Museum of Art, New York, gift of Emily Fisher Landau in honour of Tom Armstrong. © ARS, NY and DACS, London 2002.

FIGURE 5: The Metropolitan Museum of Art, New York, gift of Mrs Alma Wertheim, 1928. (28.130.2).

FIGURE 6: National Portrait Gallery, Smithsonian Institution Photograph © 1997 Estate of Paul Colin ADAGP, Paris and DACS, London 2002.

CHAPTER 8

FIGURE 1: Photo by W. Leitner

FIGURE 2: Photo by H. Simon.

FIGURE 3: Photo by H. Maurer.

FIGURE 4: Photo by Römisch-Germanisches Zentralmuseum Mainz.

FIGURE 5: Photo by Römisch-Germanisches Zentralmuseum Mainz.

FIGURE 6: Photo by Römisch-Germanisches Zentralmuseum Mainz.

FIGURE 7: Photo by Römisch-Germanisches Zentralmuseum Mainz.

# Index